Progressive Engineering Materials

HODDER AND STOUGHTON
LONDON SYDNEY AUCKLAND TORONTO

Cover photograph Courtesy of Rolls-Royce plc

ISBN 0–340–36435–1

First published 1988

Typeset by Macmillan India Ltd, Bangalore 25
Printed in Great Britain for Hodder and Stoughton Educational,
a division of Hodder and Stoughton Ltd, Mill Road, Dunton Green,
Sevenoaks, Kent by Page Bros, Norwich

Contents

Preface   vi

Introduction   vii

Chapter 1   *States of matter*   1
1.1   Changing states of matter   1
1.2   States of matter   1
1.3   Bonding between atoms and molecules   3
1.4   Ionic bond   5
1.5   Covalent bond   5
1.6   Metallic bond   7

Chapter 2   *Atomic structure of metals*   9
2.1   Solidification process   9
2.2   Unit cell and space lattice formation   11
2.3   Typical unit cell formation   12
2.4   Dendritic growth   18
2.5   Definitions of properties   21
2.6   Influence of structural formation on mechanical properties   23
2.7   Grains and grain boundaries   24
2.8   Crystalline definition   29
2.9   Crystalline materials   30

Chapter 3   *Binary thermal equilibrium diagrams*   31
3.1   Equilibrium conditions in alloys   31
3.2   Equilibrium cooling   32
3.3   Construction of equilibrium diagrams   33
3.4   Definitions and terminology of equilibrium diagrams   34
3.5   Types of equilibrium diagrams   36
3.6   Complete solid solubility equilibrium diagram   39
3.7   Partial solid solubility equilibrium diagram   40
3.8   Purpose of binary equilibrium diagrams   43

Chapter 4   *Coring and precipitation process*   44
4.1   Influence of industrial cooling rates   44
4.2   Definition of coring   44
4.3   Mechanism of coring   46
4.4   Elimination of coring   47
4.5   Definition of precipitation   49
4.6   Mechanism of precipitation hardening   51
4.7   Advantages and limitations of precipitation   53

Chapter 5   *Iron–carbon alloys*   54
5.1   Composition and structure   54
5.2   Structure and properties   61
5.3   Iron–carbon equilibrium diagram   65
5.4   Definition of constituents—ferrite, austenite, cementite and pearlite   67
5.5   Iron–carbon alloys and heat treatment   68
5.6   Annealing processes   69
5.7   Normalising   73

5.8	Quench hardening	73
5.9	Critical cooling rate	74
5.10	Tempering processes	75
5.11	Nitriding	76
5.12	Mass effect and hardenability	76
5.13	Time–temperature transformation curve construction	77
5.14	Martempering	78
5.15	Austempering	78
Chapter 6	*Heat treatment equipment and procedures*	82
6.1	Liquid bath furnace and applications	82
6.2	Muffle furnace and applications	84
6.3	Non-muffle furnace and applications	84
6.4	Gas furnace atmospheres	87
6.5	Quenching media	88
6.6	Quenching procedure	89
6.7	Safety precautions	89
Chapter 7	*Cast irons*	91
7.1	Definition of cast iron	91
7.2	Cast iron and the iron–graphite thermal equilibrium diagram	92
7.3	Cooling and microstructure of hypoeutectic irons	93
7.4	Effect of cooling rate and section variation on structure	94
7.5	Influence of composition on structure	96
7.6	Ordinary cast iron—grey and white	98
7.7	Malleable cast iron—whiteheart, blackheart and pearlitic	101
7.8	Inoculated cast iron—pearlitic high duty and spheroidal graphite	105
Chapter 8	*Light alloys*	108
8.1	Heat treatment of light alloys	108
8.2	Pure aluminium	109
8.3	Elements alloyed with aluminium and their effects	110
8.4	Heat treatment of aluminium alloys	111
8.5	Aluminium alloys	114
8.6	Pure magnesium	120
8.7	Elements alloyed with magnesium and their effects	124
8.8	Heat treatment of magnesium alloys	127
8.9	Magnesium alloys	127
Chapter 9	*Copper and zinc alloys*	134
9.1	Pure copper	134
9.2	Commercial grades of copper	134
9.3	Elements alloyed with copper and their effects	135
9.4	Common copper alloys—brasses, tin bronzes, aluminium bronzes, cupro-nickels and nickel-silvers	139
9.5	Pure zinc	147
9.6	Zinc alloys and their applications	149
Chapter 10	*Molecular structures of polymers*	152
10.1	Introduction to polymers	152
10.2	The role of carbon in polymer structures	152
10.3	Terminology in polymer technology	155
10.4	Common molecular chain arrangements	157

Chapter 11   *Effect of structure and additives on properties*   161
     11.1   Types of structures in thermoplastic polymers   161
     11.2   The effect of structure on mechanical properties   164
     11.3   Additives in polymers, their purpose and application   165

Chapter 12   *Common polymeric materials*   169
     12.1   Polyethylene   169
     12.2   Polyvinyl chloride   172
     12.3   Polystyrene   174
     12.4   Polyamides (nylons)   177

Chapter 13   *Testing of materials*   188
     13.1   Reasons for testing   188
     13.2   Hardness testing   189
     13.3   Bend tests   189
     13.4   Standardisation of materials testing methods   191
     13.5   Tensile testing of materials   195
     13.6   Torsion testing   207
     13.7   Hardness measurement   208
     13.8   Impact testing   215

Chapter 14   *Corrosion*   219
     14.1   Introduction to corrosion   219
     14.2   Theory of corrosion   219
     14.3   Surface oxidation of metals   220
     14.4   Electro-chemical corrosion   220
     14.5   Potential difference in corrosion cells   223
     14.6   Differential aeration corrosion   225
     14.7   Composition and structure of metals   226
     14.8   Environmental conditions   228
     14.9   Structural design   228
     14.10   Stressing and temperature conditions   229
     14.11   Summary of corrosion prevention   229
     14.12   Corrosion protection   229
     14.13   Corrosion-resistant alloys   229
     14.14   Cathodic protection   230
     14.15   Protective coatings   231
     14.16   Corrosion inhibitor chemicals   231

Chapter 15   *Factors affecting materials selection*   233
     15.1   Designer's considerations   233
     15.2   Mechanical properties   233
     15.3   Component cost   233
     15.4   Service requirements   234
     15.5   Functional requirements   235
     15.6   Material properties   236
     15.7   Material structure   238
     15.8   Material utilisation   241

Index   246

Preface

This textbook has been designed to cover the principal learning objectives of the standard BTEC units for Engineering Materials Technology at Levels 2 and 3 in the A5 Programme for Mechanical and Production Engineering. The technology of engineering materials will not have been studied previously and the level 2 Unit is pre-requisite for the standard unit of Manufacturing Technology 3. Therefore the authors have included supplementary material that they consider important to give a better understanding of the subject.

It is hoped that this book will be of use to students other than those involved in BTEC studies as a reference book for further study on the subject of materials.

The authors would like to thank everyone who has been of assistance in any way in the preparation of the text. In particular we should like to record our appreciation to the companies who have provided valuable information and photographs—essential in a technical book of this nature. Finally, our sincere thanks go to our wives who have been so patient and understanding and who have given us such support and encouragement.

Introduction

Several thousand years ago the most important material to man was flint, which was used as a cutting material in the form of crude knives, axes, spears and arrows. Although we consider the implements of ancient man as crude, at the time of their use they represented the most advanced technological developments: they were certainly a vast improvement on the use of animal bones as tool materials. Inevitably, by accident, man found usable pieces of metallic gold, copper and silver, the three principal metals that occur in the uncombined or *native* condition. The use of such materials is highlighted by archeological discoveries of jewellery items. Ornamental use of these materials proved very satisfactory because they were a pleasant colour and could be easily crafted.

The other source of materials was 'heaven'—in the form of meteorites which supplied man with an iron-nickel alloy. Early man, however, did not realise the nature of the alloy at that time. In fact, metallography, or the study of metals and their behaviour, is a comparatively recent science.

The first smelting of metallic ore would have been by accident. A piece of ore was probably included as a fire-surround stone and owing to the prolonged heat the metal separated out from its ore to remain as residue in the fire. Ultimately, a use for this strange material was found. One such material we would describe as wrought iron. From some of the ancient paintings historians have concluded that early man considered all of the earth's substances to be alive and to comprise various amounts of fire, earth, water and air. Copper was thought to be made from earth, water and large quantities of fire which provided copper's rich red colour. This theory was adopted by Aristotle and existed until 500 years ago.

Democritus, another Greek philosopher, provided an atomic theory of substances and he believed the Universe to be composed of particles and empty space. This model provided a very crude foundation for our present atomic appreciation of matter. Two thousand years ago there was no experimental work as we know it today, but man had learned to cast intricate shapes, to forge simple artefacts and also to increase the hardness of iron, by a process described today as *cementation*. Early man did not know that this process was the result of adding carbon to iron. Man also knew of quench-hardening and tempering procedures but did not appreciate why such operations gave the results they did. The outcome of the tempering process was considered to be the result of relieving evil spirits from within the metal.

It was not until the seventeenth century that through experimental science rapid advances in the theory of materials and their structures were made. The earliest developed alloys were copper-based and, because they resembled gold in colour, they were used for jewellery. It is possible that craftsmen involved in producing such alloys actually thought that they had produced gold.

The research to produce special materials for solving current engineering problems is very intense. The origins of the metallurgical research, testing and application probably lie in the work of alchemists who mixed, brewed and distilled in vain in their search for desirable substances that would change base metal into gold. The present scientific theories and practices of materials science and materials testing have made phenomenal progress from the ancient practice of alchemy.

We can now show that all matter is composed of atoms. The atom has a nucleus which contains protons and neutrons. The proton is described as having a positive charge and the neutron, as its name suggests, has no charge. In 1913 the Danish physicist Niels Bohr suggested that electrons orbited around the nucleus in a manner similar to the orbitting of planets in our own solar system. In the early part of the last century, the electron was considered to have a positive charge, but it has now been shown that the charge is, in fact, negative.

As more research is conducted on atomic interactions, our knowledge of materials and their structures will be greatly improved. Scientists, physicists and metallurgists have been able to provide a model of the spatial relationships between electrons and the nucleus. Such a model suggests that if the nucleus of the atom is represented by a tennis ball then the nearest electron, represented by a pea, would be positioned some 75 metres from the nucleus. This model gives an indication of the minute size of the atom and the relatively large amount of space between electrons and atomic nuclei.

The current interest in the atom is as a source of energy which will help to relieve dependance upon fossil fuels. Atomic energy has many advantages, the principal one being that there are substantial sources of fuel. In atomic energy research, problems of excessive heat, corrosion and radiation hazards have resulted in the development of new materials and alloys.

One of the results of the world fuel crisis, is that engineers have had to produce components that are as light as possible and yet still retain their strength. This improved weight-to-strength ratio has involved the development, production and application of plastics as well as new alloys. This book outlines the basic principles involved in material structures and the appropriate processes and treatments that can improve the properties of metallic and non-metallic materials.

Chapter 1

States of matter

1.1 Changing states of matter

Atoms may be considered as the building blocks of matter. The way the atoms combine with each other and the distances maintained between them will govern the state or condition of the matter. The terms *solid*, *liquid* and *gas* are used to describe the various conditions of matter. It is scientific practice to refer to the melting or solidification point and the boiling or condensation point when describing a substance or a quantity of matter that has undergone structural changes with respect to temperature changes.

Certain atoms may group together to form a specific material; but the same atoms may group together in a variety of forms. For example, carbon atoms can group to form two crystalline substances, namely graphite and diamond. A slightly different example would be the grouping of the H_2O molecules at different temperatures. Consider what happens when an ice cube is held in the hand. It absorbs heat energy from the hand and melts to form water. Holding the wet hand in front of the fire enables further heat energy to be absorbed. This causes the water to evaporate as vapour (a condition which may be considered as gaseous).

Ice (solid), water (liquid) and water vapour (gas) are examples of these common states of matter. Whilst undergoing these changes of state the chemical composition of the substance does not change. The ice, water and its vapour are all composed of H_2O molecules. If the chemical composition remains the same then what causes the change in the physical properties of the various states of matter? The solution lies with the bond force between the atoms and how such a force of attraction or repulsion alters as temperature changes.

1.2 States of matter

If heat energy is applied to a block of ice, then a change from solid to liquid will occur, followed by a change from liquid to a gas. The phases of solid, liquid and gas are the three *states* or *conditions of matter*.

If temperature and time are recorded and plotted whilst heating a block of ice, important features develop as shown in figure 1.1.

Between A and B the temperature rise is termed 'sensible' heat and shows an increase in the temperature of the ice. Between B and C the temperature rise is halted but there is still heat energy being supplied and absorbed. This is called the *latent heat of fusion*. At this stage the heat energy is used to break the bonds between the frozen water molecules which provide the geometry and solid state of

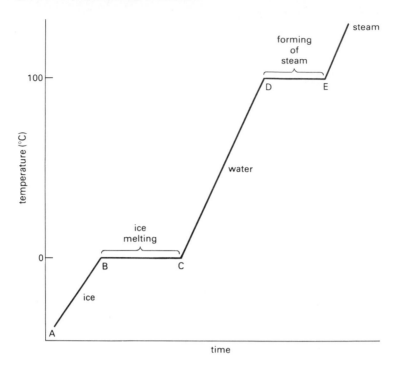

Figure 1.1 Changes in the state of matter for water with change in temperature.

the water, i.e. the ice. It takes time for all the ice to change to water as the molecular arrangement changes from *ordered* to *disordered*. This results in the forces of attraction between adjacent molecules becoming dramatically reduced, producing a liquid condition.

Between points C and D the temperature rise is again 'sensible' heat as the liquid becomes hotter. At point D the temperature rise is arrested and once again the heat energy is required to break the bonds between the molecules in the liquid. At E this has been achieved and the molecular arrangement becomes less dense with the increasing velocity given to the molecules by the heat energy input.

As the temperature increases, the velocities of the molecules also increase, resulting in an increase in volume. If, however, the volume is contained, then the increasing molecular activity will increase the pressure on the container walls.

It is the high velocity of the gas molecules that prevents them from sticking or adhering to one another when they collide, therefore the substance maintains its gaseous form. If the gas is cooled, the thermal motion of the molecules decreases and at some specific temperature when the molecules make contact with each other, the cohesive forces hold them together. The state of the substance will then return to liquid.

This close contact of the molecules makes it very difficult to compress liquids in comparison with gases, which, because of their more 'open' molecular arrangement can be compressed. The forces of attraction between the molecules give

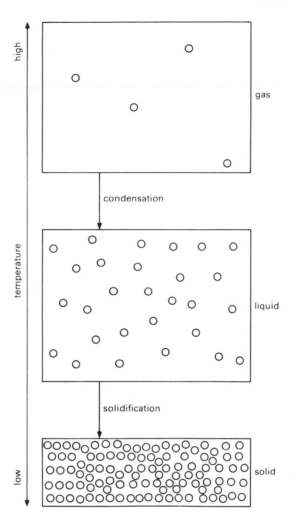

Figure 1.2 The three states of matter.

liquids a specific volume whereas a gas will expand to fill the entire volume of the vessel. If the temperature of the substance is further reduced then the molecular motion decreases until fixed positions between molecules are created. Hence a regular structure is developed and a solid is formed (figure 1.2).

1.3   Bonding between atoms and molecules

In Section 1.2 it was shown that whilst matter could have the same chemistry for its different states or forms, its physical properties changed. The atoms of hydrogen and oxygen, which produce the water molecule, assume different distances between each other when they combine to form ice, water or water vapour. The atomic forces of attraction or *bond strengths* are also different. In order to understand how these atoms come together and what keeps them in their set positions, the structure of the atom must be appreciated.

All atoms consist of a nucleus which contains uncharged or neutral particles called neutron together with positively charged particles called protons. The nucleus is surrounded by negatively charged particles known as electrons. In an electrically neutral atom there are the same number of protons in the nucleus as there are electrons surrounding it. The electrons are arranged around the nucleus in a series of 'shells', the capacity of electron in each shell not being the same (see figure 1.3 below). The first shell is full when it contains two electrons, the second shell with a complement of eight electrons whilst the third shell can hold up to 18 electrons. The fourth, fifth, sixth and seventh shells have a theoretical capacity of 32, 50, 72 and 98 electrons respectively. However, in practice, no shell contains more than 32 electrons. The number of electrons in the outermost shell is significant since it has a bearing upon the properties of the element. Some elements are extremely stable and are described as being chemically inert. They are characterised by having all their shells filled completely with electrons. With the exception of helium, which has only two electrons, inert elements have a complement of eight electrons in their outermost shell. Such elements are neon, argon, krypton, xenon and radon. All other elements, which have at least one incompletely filled shell, will take part in a chemical reaction, i.e. atoms will combine or bond with others in some way.

It is the outermost shell electrons, called valency electrons, that are responsible for the combination or bonding of elements. Interaction between atoms produces stable units called molecules. Such formation is not restricted to just two atoms. Complicated molecules can be produced almost simultaneously from reactions between several atoms or from further reactions between atoms and existing molecules. In producing a molecule the original arrangement of electrons and nuclei is disturbed and new arrangements occur. These newly formed molecules can have properties which are quite different from the constituent atom and molecules. For example, if metallic sodium and chlorine gas are brought together the result of interchanging outer-shell electrons will produce sodium chloride (common salt) which is quite different from the original constituents; the reaction that occurs when hydrogen gas is burnt in oxygen produces water vapour—H_2O!

A large number of molecules of a wide variety may be constructed by atomic interaction due to electron 'shift'. Such molecules may be *diatomic* comprising of two atoms, for example O_2 and Cl_2, or molecules involving hundreds of atoms as in organic polymers. It is also possible to have a giant molecule such as diamond or sodium chloride crystals, which can easily be seen and handled. Although there are an infinite number of molecules which may be formed, the types of bond responsible for their formation are few in number. The types of bonding arrangements can be catagorised as either primary or secondary: the bonding forces between atoms are primary forces and those between molecules are secondary forces. The primary bonds to be considered are described as *ionic*, *covalent* and *metallic*.

1.4 Ionic bond

This bond between adjacent atoms depends upon the forces of attraction created when one atom loses an electron which is then 'absorbed' by a neighbouring atom. Before losing the electron the atom is comprised of neutrons and protons in its core. The number of protons, which are positively charged, is equal to the number of negatively charged electrons. As a result of electron loss, the particle which is now described as an *ion* becomes positively charged since it contains more protons than electrons. The neighbouring atom that receives the extra electron becomes a negatively charged ion (see figure 1.3). Because of this electron 'shift' a cohesive force is developed between the newly formed positive and negative ions. Such a primary bond is described as an ionic bond.

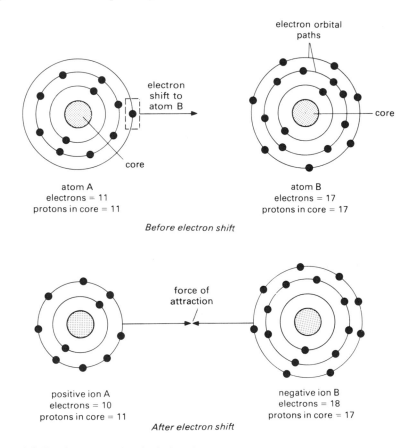

Figure 1.3 The formation of an ionic bond.

1.5 Covalent bond

Some atomic arrangements exist by sharing the electron, instead of a definite transfer of electrons from one atom to another. This type of atomic bond is described as a *covalent bond* where the shared electron enters into joint orbit around both atomic nuclei. With certain atoms, this sharing provides each nucleus with an effective complement of eight outer-shell electrons, thus forming a covalent bond (figure 1.4).

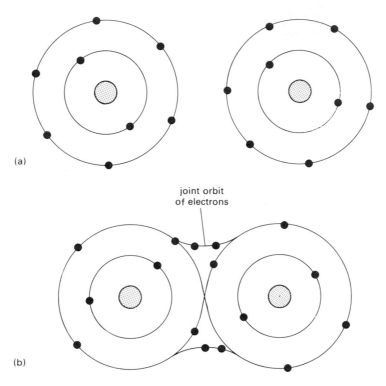

(a)

joint orbit
of electrons

(b)

Figure 1.4 The formation of a covalent bond. (a) Individual atoms. (b) Molecule formation due to cohesion force of covalent bond.

The symbol for a covalent bond is a dash, which, when drawn between two hydrogen atoms, readily indicates a covalent bond, H—H. Because hydrogen only has one electron, the 'borrowed' electron fills the first orbital shell which requires two electrons for valency of the nuclei i.e. that the shell is complete with two electrons so it becomes stable but not inert. For other elements the required number is eight electrons in their outermost shells. If carbon reacts with hydrogen in the ratio of one carbon atom to four hydrogen atoms then methane, CH_4, is obtained. This atomic arrangement is the result of covalent bonding: the single electrons from each of the four hydrogen atoms mix with the four provided by the carbon atom, thus producing a complement of eight electrons for the carbon atom and two electrons for each of the hydrogen atoms. The eight electrons form an electron cloud arrangement. This is the basis of covalent bonding. Alternative ways of representing the methane molecule are shown in figure 1.5.

The formation of ionic bonds is the result of electron transfer whilst the formation of covalent bonding is due to electron sharing. With these two means

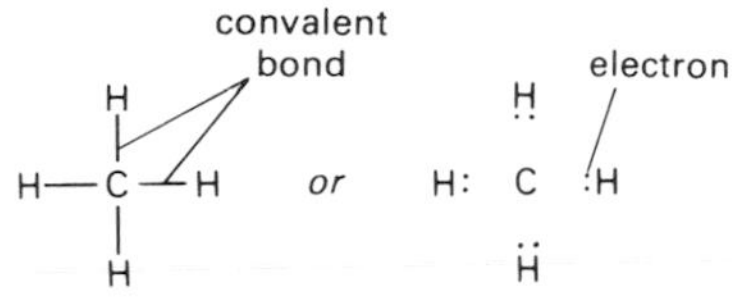

H—C—H   *or*   H: C :H

convalent bond · electron

Figure 1.5 The atomic arrangement of methane.

of bonding the outcome is the provision of eight electrons in the outer shell of each atomic nucleus, excluding of course, the hydrogen molecule, which requires two electrons.

1.6 Metallic bond

This bond differs from the ionic and covalent bonds in that the electrons of the metal ions 'wander' around the lattice arrangement of the metal and are continually involved with more than one ion. This random motion of electrons in the electron cloud is shown in figure 1.6.

The metallic bond is not strictly confined to one type of metal ion, i.e. pure metals. It will, in fact, develop between different sized ions and hence different metals can be combined by this bond. Such combinations are the basis of alloy formations such as brass alloys produced from copper and tin.

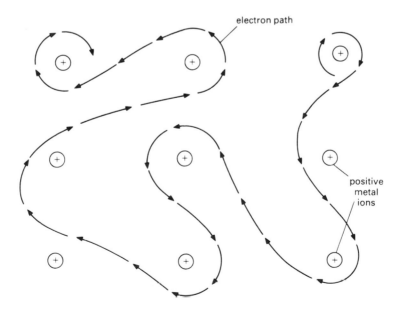

electron path

positive metal ions

Figure 1.6 Metallic bond formation by random electron motion. The possible path of a single electron between metal ions is shown.

The ease with which the electrons move between the metal ions is also responsible for the electrical conductivity properties of the majority of metals.

In the solid state, an assembly of numerous ions, either positive or negative will form a symmetrical geometric arrangement or *crystal*. Some common crystal formations such as body centred and face centred cubic structures are discussed further in Chapter 2.

Chapter 2

Atomic structure of metals

2.1 Solidification process

The change in a metal from a solid to a liquid condition is achieved by the increased vibration activity of the atomic structure as the temperature of the metal increases. The vibrations become vigorous enough to break the bonds; the atoms loose their ordered geometric positions and start moving at high velocities in all directions. The metal is now in a liquid state and in this condition there still exist very weak forces of attraction between the atoms as shown in figure 2.1.

When the temperature of the molten metal starts to fall, the velocity of the moving atoms begins to decrease, i.e. the kinetic energy of the atoms decreases. As

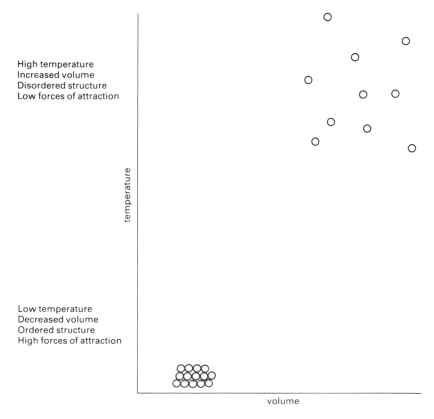

Figure 2.1 The effect of temperature change on an atomic arrangement.

9

the random motion of the atoms becomes sluggish due to heat loss, contact between some atoms will occur. This contact will result in the formation of nuclei around which metal crystals will eventually form. This type of nucleus should not be confused with an atomic nucleus. However, both uses of the word refer to the centre of a structure. The growth of the crystals will render the liquid more viscous and solidification can be considered to have started.

During the solidification process the atomic formations develop and grow in particular ordered arrangements but in random directions. The ordered patterns are characteristic of a particular metal. These patterns are discussed in detail in Section 2.3 — Typical unit cell formations.

If the cooling rate of a pure metal is observed over a period of time, then a distinct *temperature arrest point* is noted. This arrest point is characterised by a horizontal line as shown in figure 2.2.

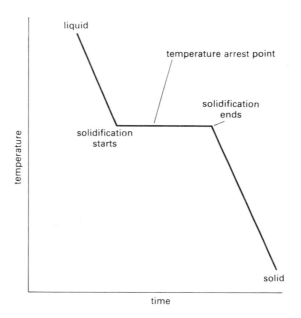

Figure 2.2 The cooling curve for pure metal.

Between the arrest points there is no fall in temperature but heat energy continues to be given out. This heat energy is described as *latent heat* and is produced from the decreasing random velocity of the atoms during solidification, i.e. kinetic energy converted to heat energy (figure 2.2). The heat energy evolved balances the heat lost by radiation from the pure metal, hence the horizontal line on the graph. Certain alloy compositions also produce a similar cooling curve. When random atomic motion has ceased, so that no liquid remains, then solidification is complete and no more latent heat is evolved, the temperature of the solid continues to fall to the ambient temperature.

The latent heat evolved during solidification may also be considered to be the heat energy required to break the metallic bonds of the pure metal. These bonds may be taken as the forces which hold the atoms in a regular geometric arrangement. For most metals the final volume after solidification is smaller than the initial volume of the liquid because the atomic arrangement of the solid is ordered, whereas with a liquid there is no fixed atomic arrangement. The shrinkage associated with solidification can be seen in figure 2.3 which represents the top of a cast ingot.

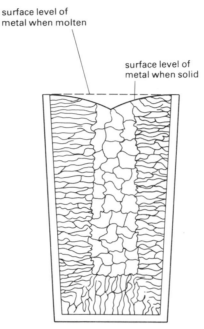

Figure 2.3 A cast ingot showing volume reduction on solidifying.

Pure metals possess certain desirable properties such as corrosion resistance or electrical conductivity. If, however, properties of high tensile strength or high degrees of hardness are required then other elements may have to be added to the pure metal. The introduction of other elements to a pure metal is described as *alloying*.

When a metal consists of two elements it is described as a *binary* alloy, 'bi' indicating two. The presence of two elements will affect the shape of the cooling curve. Since the molten alloy now contains atoms of different size, the kinetic energy of the atoms will be different. The heat energy evolved will therefore have different values which will produce a solidification-temperature gradient. An alloy or non-uniform atomic material will solidify over a temperature range as illustrated in figure 2.4.

2.2   Unit cell and space lattice formation

The fact that metals become molten when heated and solid when cooled is useful to the engineer, since it allows greater versatility of application for most metals.

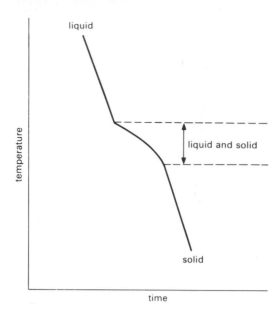

Figure 2.4 The cooling curve for a binary alloy.

For casting operations it is necessary for the metal to be in the molten state whilst for forging and cold-working operations the metal is required in the solid state. The solid state is achieved by the formation of crystals or grains as molten metal cools. Each crystal, if examined with a high-powered microscope called a field ion microscope, reveals evidence of distinct atomic patterns peculiar to that type of metal. Figure 2.5 is a photograph of the hexagonal atomic arrangement of a titanium specimen taken through such a microscope.

The initial formation or grouping of atoms to produce the first atomic patterns during solidification is described as a *unit cell*. Many such unit cells are produced simultaneously at the start of solidification and are considered to be the nuclei for the eventual crystal formation. Unit cells identified as body-centred cubic, face-centred cubic and close-packed hexagonal are illustrated in figures 2.6 to 2.8.

As solidification continues more atoms position themselves around the unit cells to reproduce a particular atomic pattern or arrangement. Such a collective growth of unit cells is called a *space lattice*. The unit cells and space-lattice arrangement have a direct bearing upon certain properties such as malleability. Certain atomic arrangements are more malleable than others. Typical space-lattice arrangements developed from unit cells are shown in figures 2.9 to 2.11.

2.3   Typical unit cell formations

Due to the development of X-ray analysis and field ion microscopes, metallic structures can now be examined to such a magnification that the ordered atomic arrangements previously referred to can actually be seen (figures 2.5 and 2.12).

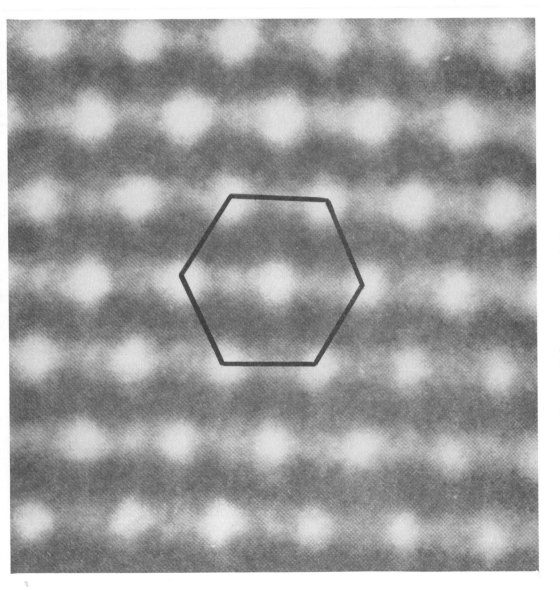

Figure 2.5 Atomic arrangement in a titanium specimen magnified 2.5 million times. The hexagonal atomic structure can be clearly seen.

Figure 2.6 Spheres arranged to represent atoms in a typical body-centred cubic unit cell.

Figure 2.7 Spheres arranged to represent atoms in a typical face-centred cubic unit cell.

Figure 2.8 Spheres arranged to represent atoms in a typical close-packed hexagonal unit cell.

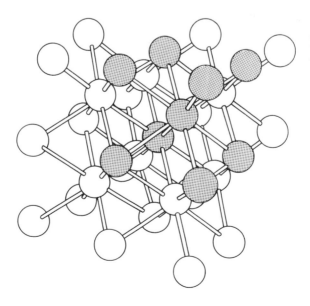

Figure 2.9 The space lattice developed from the body-centred unit cell. The unit cell is indicated by the shaded spheres.

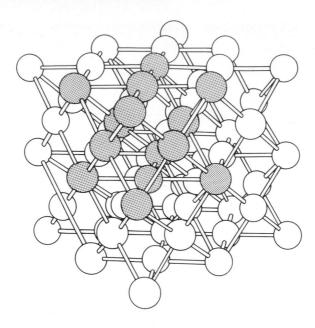

Figure 2.10 The space lattice developed from the face-centred unit cell. The unit cell is indicated by the shaded spheres.

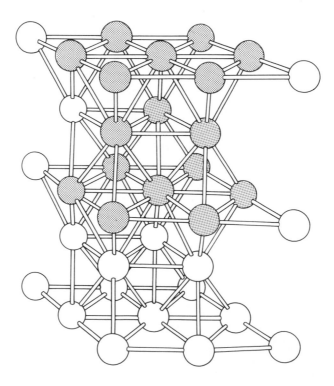

Figure 2.11 The space lattice developed from the close-packed hexagonal unit cell. The unit cell is indicated by the shaded spheres.

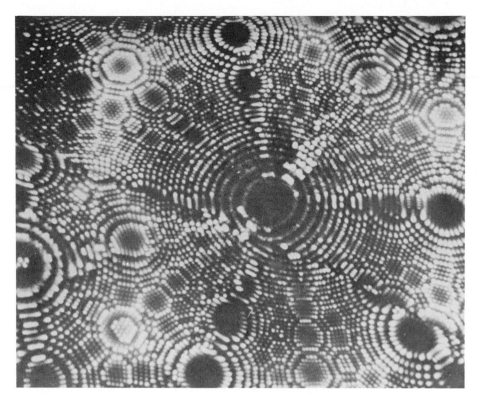

Figure 2.12 The presence of an ordered atomic arrangement can be clearly seen in the micrograph of the tip of a tungsten needle. (Magnification X 5 000 000).

Before such techniques were developed, calculated assumptions were made about the structure of metals. One such assumption was that the structure consisted of disordered molecular arrangements. Modern metallurgical analysis has shown that for true metals there exist three basic structures, namely body-centred cubic (BCC), face-centred cubic (FCC) and close-packed hexagonal (CPH) (figures 2.6 to 2.8). Certain metals change their atomic arrangement with changes in temperature. An example of this is iron which at room temperature is BCC but at around 900°C changes to FCC. Chapter 5 deals with the properties and structures of ferrous alloys.

To assist understanding of the atomic structures, models are used in which atoms are considered as spheres. If the centres of the spheres lie at the corners of a cube and each sphere touches the surface of its neighbour so that the length of one side of the cube is equal to the diameter of the sphere, then a simple cubic structure exists. This structure contains eight spheres. Such a structure is not associated with engineering metals but is used as a datum for the more common atomic arrangements.

By positioning an extra sphere into the centre of the simple cubic cell, a BCC structure is produced (figure 2.6). This structure contains nine spheres. When spheres are located at the corners and at the centre of each of the six faces of a

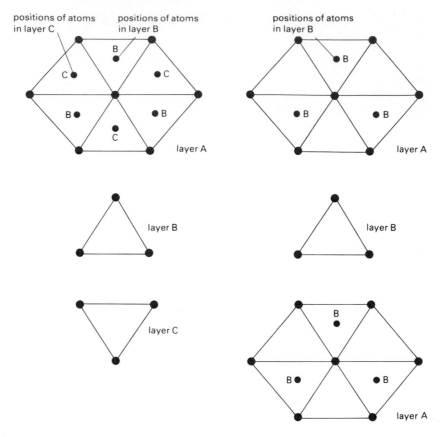

Figure 2.13 Close-packed lattice structures.

simple cube then a FCC structure is produced (figure 2.7). This structure contains 14 spheres. A structure described as CPH is obtained by alternate hexagonal and triangular atomic formations. The hexagonal layer has six atoms surrounding a single atom. Below this hexagonal formation are three atoms in the form of a triangle beneath which is another hexagonal layer to complete the unit cell (figure 2.8). The FCC and the CPH structures are observed as alternative ways of close atomic positioning or packing. The difference between the two arrangements is the position of the horizontal planes of the triangular arrangement of atoms relative to the hexagonal plane, as shown in figure 2.13.

2.4   Dendritic growth

As outlined in Section 2.1, when solidification begins, a nucleus is formed by atoms. Depending upon the atomic unit cell arrangement of the metal, *growth* or *atomic build-up* occurs in certain preferred directions. These preferred growth directions produce atomic arrangements or patterns that have a pointed or spiked appearance and such arrangements of atoms are described as *dendrites*.

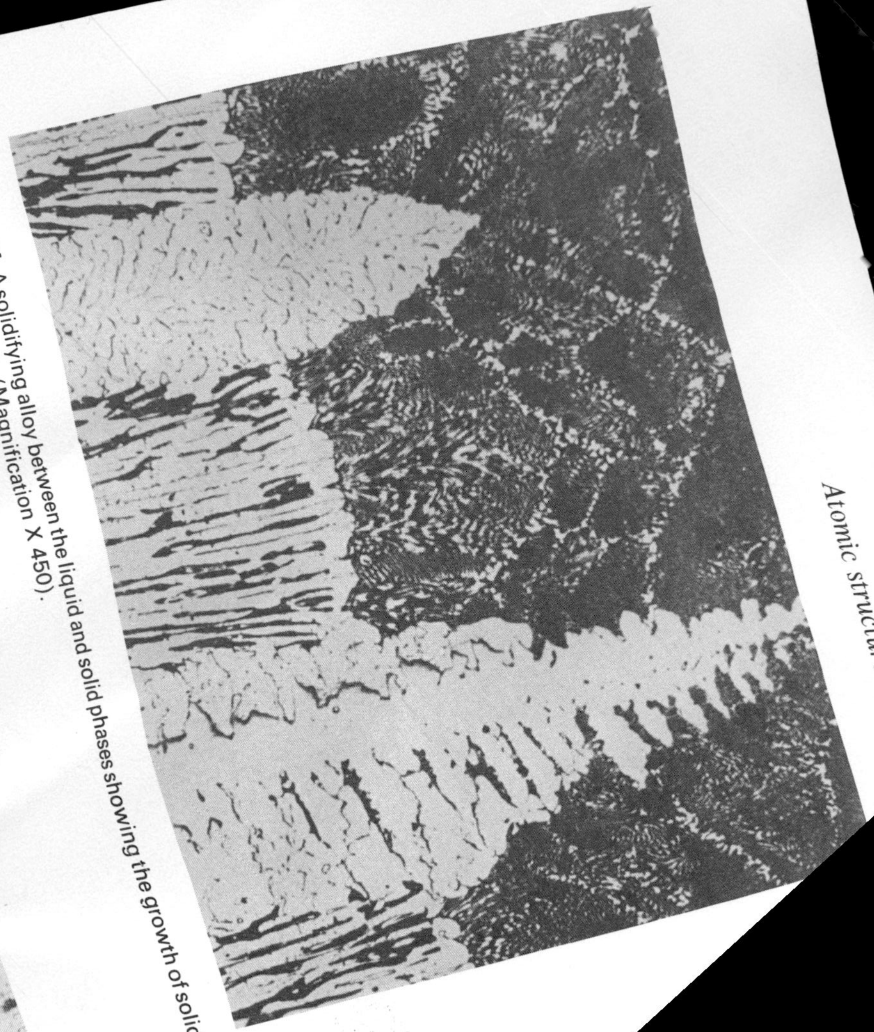

Figure 2.14 A solidifying alloy between the liquid and solid phases showing the growth of solid dendrites. (Magnification X 450).

Figure 2.15 Antimony showing the dendritic formation of the surface

of the dendrite arms. This changing composition will be

4.

of properties

ustrial terminology has produced or been responsible for
ases which are used to describe people or articles. Typical
l as nails', 'tough as cast iron' or 'strong as cast iron'.
rases are quite misleading to the engineering student who
n is anything but tough; it is in fact quite brittle. The most
f materials will now be considered in detail.

ribed as being strong has a structure that remains intact or
ted to a comparatively large load. The atomic or molecular
withstanding the external load without appreciable *atomic*
g person is able to overcome the resistance created by the
wever, a weak person is unable to overcome the resistance
an result in strain or even slip within the human structure
scle or a slipped disc. This human example highlights the
interactions which occur due to external loading dictate
man or otherwise, metallic or non-metallic, is described
o be realised that the strength of a material
ome materials are strong when load
other directions. For example, pla
ion, therefore they are widely used in
ve stresses are encountered. Unfortu
steel reinforcements have to be used t
ithout which the concrete would fail

material to return to its original dimensio
noved. The property depends upon the at
legree of elasticity is due to the magnitud
en the atoms when an external force tries to
als exhibit different degrees of elasticity. C
ped as *elastomers* (see Section 10.3) have
whilst other materials such as glass or ceramic

...is a tree.
...cation process is very important because
...trols the rate of solidification. A spike formation has a
...timony—tree-like in
...surface area-to-volume, higher than a spherical formation, and
...solidification growth since the areas of the metal subjected to this release of heat

therefore more heat can be dissipated thus maintaining the solidification process.
During solidification of metals, latent heat is evolved which is detrimental to
deter dendritic growth since the areas of the metal subjected to this release of heat
branching in more favourable directions occurs producing a network of arms.
The dendrite formation can be observed on partially frozen surfaces of water or
even on a piece of galvanised metal.

The growth of dendrites occurs instantaneously and is three-dimensional. This
development continues until an obstacle in the form of an adjacent dendritic
growth pattern or the walls of the mould prevents further progress in that
direction. The growth directions of each dendritic arrangement are random, and
therefore each evolving grain will have specific directional properties i.e. when
stressed the grains will deform or slip more easily in certain directions than in
others. Hence, each grain will possess individual 'directional' properties as shown
in figure 2.16. The grain or crystal orientation is discussed in Section 2.7.

When all the liquid has been absorbed by the ever-growing dendrites (see
figure 2.14) the grains are fully formed and the material is solid.

The dendrite formation or growth pattern is very difficult to detect in some
pure metals but in an alloy it is more obvious. This is because a range of
temperatures exist during solidification (see figure 2.4) in conjunction with the

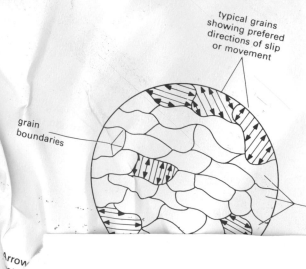

typical grains
showing prefered
directions of slip
or movement

grain
boundaries

grains

Arrow...
...ate

shape after the force has been removed. Such a property is required in stamping
and forging operations such as the manufacture of coins and medals where the
impression must remain on the metal.

Ductility

A material that can elongate easily without fracturing when loaded in tension is
described as being ductile. Copper materials are very ductile. This property is
utilised in the process of producing very fine wire by drawing down the copper
from a larger copper section. It will be appreciated that a dimensional change must
occur when a tensile loading situation arises, the material reducing in diameter
whilst extending in length. Ductile materials are also capable of resisting fracture
whilst atomic slip takes place. If subjected to shock loading the material would
yield and eventually deform.

Malleability

This property must not be confused with ductility. Malleable materials are easily
formed without fracture of the structure occurring. However, such materials are
not capable of being drawn out into wire. Consider a piece of lead which can be
beaten and rolled into sheet form but it cannot be pulled through a *die* to produce
wire. It can be *extruded* into wire form if required. To understand this, consider
toothpaste which can be formed into a ribbon by extrusion from a tube—a simple
operation and highly successful. However, if the toothpaste were to be pulled into
ribbon form then the exercise would obviously fail. This analogy is applicable to
lead rod production; lead can be forced through a die—extruded—but it cannot
be successfully drawn.

Brittleness

A brittle material has a tendency to fracture rapidly when subjected to a shock
load or sudden blow. There is very little atomic slip or permanent deformation
before fracture. Glass at room temperature is an example of a brittle material. The
temperature of the component can affect the degree of brittleness. Materials at
very low temperatures act in a brittle manner since their atomic structure cannot
slip or move fast enough to absorb the sudden load; the mobility of the atomic
structure is dependant upon temperature. High temperatures in materials
produce high atomic mobility whereas low temperatures produce very low rates
of atomic movement. The higher the degree of atomic mobility, the easier it is for a
material to absorb sudden loading by allowing the atoms to slip. Thus the
material deforms as opposed to fracturing. The hot-forging processes rely on slip
and deformation taking place, rather than sudden fracture, when the material is
worked.

Toughness

This is considered to be the opposite to brittleness and it is the ability of a
material to absorb sudden loading without fracture...

temperature and is a property that is often desired in conjunction with hardness. A typical example of the combined properties of toughness and hardness is a correctly hardened and tempered cold chisel. The hardness allows the cutting edge to function satisfactorily whilst a toughened shank allows hammer blows to be absorbed without fracture.

Hardness

The property of hardness is the resistance of a material to abrasion or indentation. Resistance to abrasion may be considered as a function of service (wear) whereas resistance to indentation may be considered a function of test. A material to be used as a cutting tool is required to possess the property of hardness. This will ensure that the cutting edge and the tool face are capable of resisting abrasion (wear), so maintaining good cutting qualities and a long life between regrinds. Another example where hardness is required is in slip gauges. Their accuracy for precise measurement depends upon no wear taking place on the measuring faces during their working life. To ensure that this accuracy is maintained the slip gauges are hardened by a heat-treatment process. Heat treatment is discussed in detail in Chapter 5. Resistance to indentation is the principle on which many hardness-testing machines are based. Such machines assess the hardness of a material by the extent of indentation of a standard indentor under a standard load on the surface of the material. A small indentation results from high resistance and so indicates a hard material.

2.6 Influence of structural formation on mechanical properties

Metals of a single type of atom are called *pure* whereas those of mixed types of atoms are called *alloys*. Properties of metals depend not only on the type of constituent atoms but also on how the atoms are assembled or arranged.

The basic atomic arrangement or pattern is not apparent in the final component such as a car door panel or a nut and bolt, but the properties of the individual crystals within the metallic component, which are controlled by their atomic arrangement, are responsible for their usefulness in engineering.

The strength of a piece of metal is determined by its ability to withstand external loading. The structure of the metal responds internally to the applied load by trying to counteract the magnitude of the applied load and therefore tries to keep the constituent atoms in their ordered positions. If, however, the load is high enough then the force which holds the atoms in place, described as the metallic bond, becomes ineffective and the atoms are then forced into new positions. This movement of the atoms is called *slip*. When performing a forging operation the material only deforms because movement or slip of the atoms has occurred on a large scale. If a material is loaded in tension the extension of the specimen will not take place if the atoms do not slip. By the same token, in hardness testing of materials using an indenter, the technique is based on the principle of forcing atoms into positions that will produce an impression in the material. The ease with which the atoms move or slip is an indication of the hardness. With the appropriate load, large slip movement would produce a large impression and indicate a soft material whereas a small indentation or impression would indicate a hard material.

The relative movement of atoms or slip within a material has a direct bearing on the mechanical properties of a material. Copper, considered to be reasonably ductile over a range of temperatures, is a material which can be drawn or extended without fracture. It is this property of ductility which allows copper billets to be eventually drawn down to a fine wire. Copper consists of a face-centred cubic arrangement of atoms and it is this geometric pattern which is responsible for the relative ease with which copper atoms can be displaced. The atomic arrangement in the face-centred structure consists of close-packed planes (see figure 2.7). Generally speaking, the greater the density of atoms within a certain close-packed plane, the greater the spacing between adjacent layers. The further apart the layers are, the weaker is the bond between these layers. Hence, bonds between close-packed planes are relatively weak and such a geometric arrangement leads to ease of slip or *deformation*.

If attempts are made to forge a piece of unheated steel, great amounts of energy are required for very small amounts of deformation. In such conditions, the material would probably rupture before the desired shape had been obtained. However, the same piece of steel, when heated to a temperature of 910°C, would readily deform. The reason for the improved forging qualities is that the BCC atomic arrangement within the cold material re-arranges to form a FCC structure which slips more easily. Therefore steel is a stronger material at low temperatures when it has a BCC atomic arrangement. The FCC arrangement of atoms is associated with gold and lead. Lead quickly and easily alters its shape when acted upon by a force and the ease with which gold will spread without rupture to produce the very fine gold leaf is due to its face-centred atomic arrangement.

The ability of metals with a CPH atomic arrangement, to deform by atomic slip is more restricted than for FCC structures. As previously mentioned, the denser the atomic arrangement, the easier slip will occur. A dense or close-packed arrangement occurs in the top and base of each hexagonal formation. The CPH atomic arrangement contains one set of close-packed planes per unit cell whereas with the FCC arrangement there are four sets of close-packed planes and hence a greater tendency for slip to occur. Structures with a BCC arrangement possess no close-packed planes and therefore the distance between adjacent layers is small, resulting in greater atomic attraction. In such an atomic arrangement greater force is required to initiate atomic slip. Hence, BCC structures are stronger than FCC or CPH structures.

2.7   Grains and grain boundaries

The process of solidification in which the development and growth of particular atomic patterns which produce formations described as dendrites has already been discussed. The growth rate and growth directions are highly variable with the result that the grains which are eventually formed from the dendrites will be randomly orientated to each other. The amount of growth from each dendritic development during solidification is restricted by the growing pattern of adjacent dendrites, the degree of restriction having an influence on the final grain size. If there are only a few nuclei and hence dendritic formations, then the eventual grains will be few in number but will be large in size. If there are many nuclei there will be many dendritic developments and the production of many grains, but smaller in size. The dendrites have other restricting factors in addition to those from neighbouring dendrites. Obstacles are provided by the walls of the mould and even by the top surface of the molten metal.

In the final stage of solidification of a metal with a regular atomic arrangement, the remaining molten metal between grains attaches to the growing dendrites. Where two dendrites meet which have been growing from different directions, the regular atomic pattern breaks down. The point where the two dendrites meet at the periphery of the grain becomes the grain boundary. In figure 2.17, atoms are represented by lead shot and the regions of ordered grouping can be considered as grains. The disordered zones separating the ordered regions represent the grain boundaries.

In metallic materials each grain is joined to its neighbour at all points on its boundary, the boundary positions being dictated by the restrictions to growth of individual grains by their adjacent neighbours. Figure 2.18 shows the surface of galvanised steel and the growth pattern of the zinc. This shows clearly the areas described as grains, their random orientation and their grain boundaries.

Because the atomic arrangements of the grain boundaries are disordered and the growth directions of adjacent grains are random, the boundary areas are stressed compared with the grains. At room temperature the grain boundary material is considered stronger than the grains and therefore a fine-grained

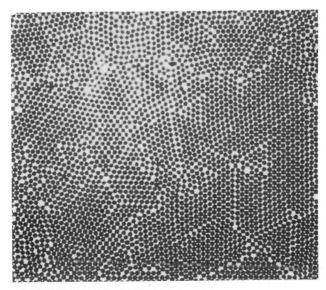

Figure 2.17 A model of grains and grain boundaries demonstrated using lead shot between two parallel plates.

Figure 2.18 Grain growth pattern as illustrated on a galvanised steel specimen.

material which contains a high degree of grain boundary area is stronger at room temperature than a structure of large grains. However, the grain boundary atoms are the last to become attached to the regular formations of the grains so they are the first to adopt random motion, i.e. melt, when the temperature of solid is increased. Therefore, for a component to operate at elevated temperatures, a more coarse-grained structure may be desirable (see figure 2.19).

An important factor associated with the solidification process and which influences the final grain size is the rate of heat loss. In the sand-casting process the grain size will be large compared with a die-cast component due to the insulating properties of the sand. The die casting solidifies more quickly due to the contact between the molten metal and the metal mould. In regions where heat loss is very rapid, such as at the face of the mould container, the formation of *super-cooled* or *chill crystals* occur. Beyond these, longer, narrower crystals develop which are described as *columnar* crystals. Their growth directions are perpendicular to the mould face and point towards the higher temperature regions of the molten metal (figure 2.19). The section thickness of the cast component also has an effect on the grain size: a thin section possesses less molten metal and hence cools more quickly (see figure 2.20).

During solidification the manner of the dendritic growth mechanism ensures that the last liquid to solidify is at the grain boundaries. This liquid usually contains a high percentage of impurities and upon completion of solidification these regions become strained due to restricted grain growth. The strained condition of the grain boundary assists the microscopic study of metals, such

Figure 2.19 A section through a lead casting showing grain formation.

areas reacting more vigorously with the chemical etching agents used in specimen preparation. Such a reaction produces a contrast between the reflected bright light from the grains and the dark image of the boundary areas. The examination of metals by optical microscopy therefore depends upon the presence and characteristics of grain boundary areas (see figure 2.21).

One may consider the grain boundary formation as a desired feature for optical surface examination. It is possible to observe certain grain structures with the naked eye or with a low-powered magnifying glass, the process being described as *macro-examination*. One such material easily observed by this method is cast zinc, a specimen of which is illustrated in figure 2.22.

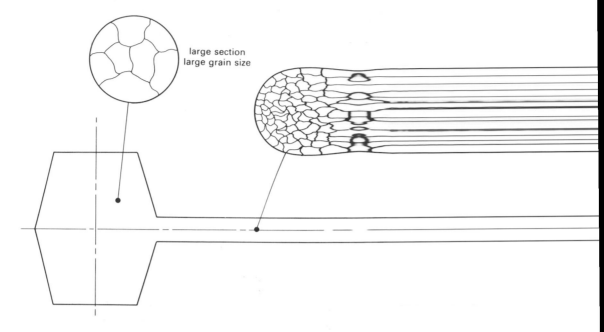

Figure 2.20 Diagram of a cast component showing the change in
cast dimension.

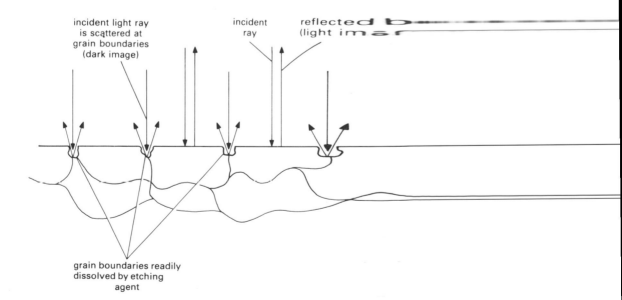

Figure 2.21 Contrasting reflected images from grains and grain bo

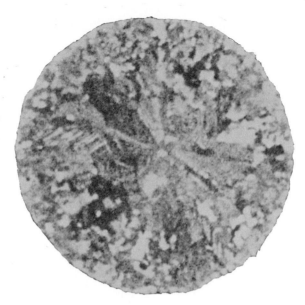

Figure 2.22 Macro-specimen of cast zinc.

2.8 Crystalline definition

Most of us are familiar with the feeling of satisfaction when the solution to a particularly difficult problem becomes 'crystal clear'. This statement implies that crystals are naturally clear. The word crystal is commonly associated with sparkling, hard materials such as crystal-cut glass or diamonds. These non-technical descriptions assume that a crystalline material should be clear, sparkling and hard. Such a description can be misleading, especially when considering engineering materials. If a material is classified as being *crystalline* the definition is based upon the atomic arrangement and not upon external appearances. It will be appreciated that many materials are relatively hard and also possess a degree of surface lustre of sparkle, but do not consist of an ordered atomic arrangement. Hence they cannot be classed as crystals. Glass is one such material. If the characteristics of lead are considered, it can be described as a relatively soft material which does not possess exceptional lustre qualities. However, it does contain a regular internal atomic structure, this being FCC at room temperature. Lead is therefore classified as a crystalline material. Gold is also a crystalline material because it has an FCC structure, but, while it has an appealing surface lustre, it is not described as a hard material. When compared with glass, gold is very soft and malleable, which highlights the possible confusion associated with the crystalline definition which relates solely to the internal atomic arrangement. All metallic engineering materials in their solid state may be described as crystalline in structure. An obvious crystalline structure can be seen in figure 2.22 which is part of a cast zinc ingot. Many engineering metallic materials require a high degree of magnification in order to observe the individual grains. The atomic pattern upon which the crystalline definition is based can only be detected by very special microscopes. One such atomic pattern is shown in figure 2.5.

In conclusion, a memorable phrase might be: 'All that shines is not crystalline'.

2.9   Crystalline materials

All metallic engineering materials are described as crystalline due to the regular geometric pattern which forms within the structure during solidification. Some metals, whilst possessing a particular atomic pattern at room temperature, are able to change to a different atomic pattern at a higher temperature. Such a mechanism of atomic re-arrangement can be of advantage to the engineer since it can improve metal-forming properties and assist in heat-treatment processes. The table in figure 2.23 gives the atomic patterns of a variety of metals at room

Body-centred cubic	Face-centred cubic	Close-packed hexagonal
iron ——— 910°C ——→ / ←——— 1400°C ——	iron	zinc
chromium	copper	magnesium
molybdenum	gold	cadmium
tungsten	cobalt ←——— 400°C —— cobalt	
beryllium ←—— 1200°C ———————————————— beryllium		
titanium ←——— 822°C ———————————————— titanium		
zirconium ←—— 865°C ———————————————— zirconium		
niobium	nickel	
vandaidum	lead	

Figure 2.23 The atomic patterns of some metals

temperature. Also shown are the metals which change their atomic patterns, the temperature at which the change occurs and the new atomic arrangement which forms. An example is iron which is BCC at room temperature and changes to FCC at 910°C and then reverts to BCC at 1400°C.

Chapter 3

Binary thermal equilibrium diagrams

3.1 Equilibrium conditions in alloys

Pure metals usually lack the desirable properties of strength and hardness which are required for many industrial applications. Consequently, ferrous and non-ferrous alloys were developed and are now used widely as engineering materials. For example steel, which is an alloy of iron and carbon, is stronger than pure iron and can also be heat treated to produce other desirable properties.

The study of alloy systems is simplified by making reference to constitutional or equilibrium diagrams. These are actually temperature–composition diagrams, which indicate the structural changes that occur during the heating and cooling of a particular alloy. Such diagrams refer to equilibrium conditions which are very rarely achieved in practice, but the diagrams provide a useful basis for the study of the alloy formation, its structural changes due to further heat treatment and the associated properties.

Alloys that contain two metals are referred to as binary alloys. Those which contain more than two metals can be assessed from the basic information obtained from the binary diagram for the two principal metals in the alloy. The constituent metals in the majority of commercial binary alloys are soluble in each other, in the liquid state or *phase*. It is possible for certain elements to produce a compound but the most common binary alloy reactions are of one of the following three types:

(1) The two metals are completely soluble in each other in the solid state— *complete solubility* type.

(2) The two metals are completely insoluble in each other in the solid state— *complete insolubility* type.

(3) The two metals are partially soluble in each other in the solid state—*partial* or *limited solubility* type.

Each of these three types of alloy produce a characteristic equilibrium diagram, which is illustrated in figure 3.5 to 3.7. These diagrams are based on a rate of cooling that is infinitely slow (however this is not achievable in real life) and which enables complete equilibrium to be maintained at all times by means of convection during the liquid phase and then by diffusion during the solid phase formation of the cooling process.

Equilibrium diagrams for binary alloys can be constructed from information obtained from the cooling curves or cooling graphs of a number of alloys

31

produced from the two metals concerned. Such cooling curves will show the characteristics related to the particular alloy as it cools from liquid to solid at room temperature.

3.2   Equilibrium cooling

Equilibrium cooling is associated with extremely slow rates of heat loss and allows unrestricted migration or flow of atoms to occur. The final crystalline arrangement will contain atoms whose positions and orientations are balanced, i.e. in equilibrium with the surrounding atomic structure. Such an atomic condition within a metal can only be achieved if considerable time is allowed for cooling to take place. Because of the huge time factors involved, equilibrium cooling is associated with laboratory projects and not production practices. In the die-casting process molten metal is poured in to the metal die and solidifies to the required shape. Solidification commences where the metal mould face and molten metal make contact and finishes within the core or centre of the mould. As with all industrial casting methods, solidification is not considered to develop under equilibrium conditions, because the time needed is not practicable for production-time requirements. A typical industrial cooling procedure is shown in figure 3.1 where position A is the region of initial solidification and position B is the last section to solidify. Because of the different cooling rates at positions A and B of figure 3.1, different crystal formations can be expected which can be clearly seen in figure 2.19. With such a variation in crystal formation there must be a corresponding effect upon properties of the material. (See Sections 2.5 and 2.6.)

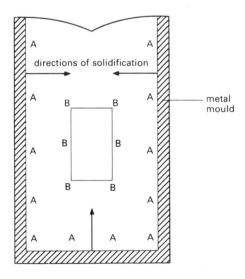

Figure 3.1 Unequal cooling rates of a typical cast ingot. A, Initial solidification regions. B, Final solidification regions.

3.3 Construction of equilibrium diagrams

In order to construct an equilibrium diagram for an alloy, it is necessary to record the cooling patterns of all the possible alloys within the range of the two metals involved. These cooling patterns, or to be more precise, time–temperature cooling curves, are prepared by monitoring every few seconds the temperature of the cooling alloy. Such detailed scrutiny of the cooling procedure will highlight any changes in the rate of cooling and any arrest points, i.e. where the temperature remains static for a period of time. Three such cooling curves A, B and C from the bismuth–cadmium alloy range are shown in figure 3.2. These curves represent pure cadmium, 60% cadmium–40% bismuth and pure bismuth.

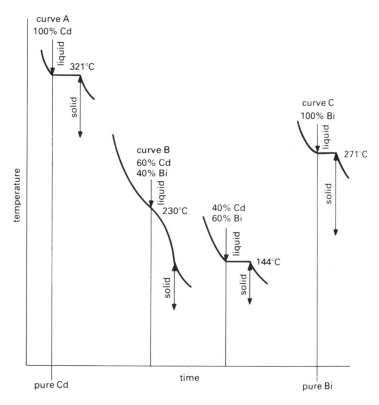

Figure 3.2 Thermal equilibrium diagram for selected cadmium–bismuth alloys.

If all the cooling curves are now positioned along a horizontal axis which represents percentage by weight of the two metals, and if the first arrest points of all the alloys are connected and likewise for the second arrest points, a diagram known as an *equilibrium diagram* emerges. Because only two metals are involved, the diagram is described as a *binary equilibrium diagram*. Since the diagram is a graphical representation, the more alloys involved, the more arrest points there are to record, and the more accurate the final diagram. A typical equilibrium diagram is shown in figure 3.3 with the cooling curves from figure 3.2 super-

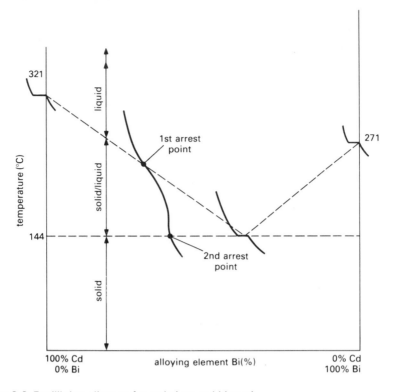

Figure 3.3 Equilibrium diagram for cadmium and bismuth.

imposed. Some interesting features of figure 3.2 are the existence of only one arrest point for the specimens corresponding to 100% cadmium and 100% bismuth which are the pure metals, and also for the alloy which produces a reaction described as a eutectic reaction. (See Section 3.4.) The alloy system of cadmium and bismuth produces one shape of binary diagram but other types exist for other alloys.

3.4   Definitions and terminology of equilibrium diagrams

When the cooling characteristics of temperature, composition and structure have been tabulated and plotted, a particular equilibrium diagram emerges for the two metals involved. Since such diagrams can provide important information about structural changes and temperature and so help in improving heat-treatment processes and casting procedures, it is necessary to provide a standardised list of scientific terminology concerning the features of equilibrium diagrams. Such standardised expressions and definitions help to reduce ambiguity when particular diagrams and their structures are being assessed.

Liquidus

At all positions above the liquidus point, the alloy is in a liquid state. During cooling, the first solid nuclei start to form upon passing the liquidus point. During heating of the alloy, the structure becomes completely liquid upon passing through the liquidus point, but just below this point some solid will still exist (see line AZD, figure 3.4).

Solidus

While cooling, the alloy becomes completely solid at the point called the solidus. During heating this is the temperature where melting of the metal begins (see line ABZED, figure 3.4).

Solvus

The solvus is a reaction point within the solid material, which denotes the amount of solubility that exists between the two metals of the alloy. This degree of solubility varies with change in temperature, i.e. as the temperature falls the amount of one element absorbed into the lattice structure of the second element decreases. When the temperature rises the degree of absorbtion increases (see slope of CB and EF, figure 3.4).

Eutectic point

At this point a reaction occurs that produces an alloy with the lowest melting temperature of the alloy range. The eutectic alloy will possess laminations of both the two solid solutions or of the two pure metals (see position Z, figure 3.4).

Solid solution

Such a material will possess unit cell formations, which will consist of atoms of both elements. As these solid solutions develop within the solid they are described by the symbols α, β, etc. These features are shown in figure 3.4 and are bounded by the lines ABC for α and DEF for β.

Eutectic reaction

This reaction produces an alloy with special characteristics. Solidification occurs at the lowest temperature for the range of alloys.

The word eutectic is derived from the Greek word *eutektikos* meaning 'capable of being easily melted'. This is indicated by position Z figure 3.4. No pasty region exists for this alloy; it possesses low melting and rapid solidification properties. When examined microscopically the structure observed consists entirely of laminations. The formation of this structure is worth detailed consideration. Before solidification commences the percentage content of each element is in harmony, e.g. 60% tin 40% lead and an equilibrium condition exists. This

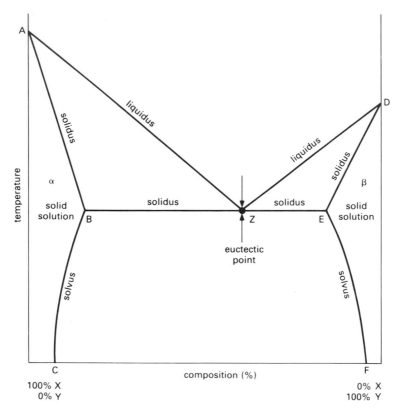

Figure 3.4 Partial solid solubility equilibrium diagram indicating specific reaction points. Liquidus: AZD. Solidus: ABZED. Solvus: BC and EF. Eutectic point: Z Solid solution α: left of ABC. Solid solution β: right of DEF.

condition can be considered similar to a beam balance with equal weights on either side of it. When solidification begins, this harmony or balance is lost the instant one of the elements forms crystal nuclei. When the first nucleus forms, efforts to regain balance are made by the second element, which itself begins to solidify. This reaction tends to 'over correct', resulting in further imbalance, and an oscillating mechanism of solidification, like a rocking of the beam balance, is created. The result of such a cooling procedure is a laminated structure. Typical alloys which respond in this manner are tin–lead, bismuth–cadmium.

3.5   Types of equilibrium diagrams

It has been shown that certain atoms are able to combine to form bonds more readily than other atoms. This readiness to produce a bond is governed by the interaction—the sharing of electrons between the atoms. In the liquid condition the strength of the bond is very low, but as the temperature falls the atoms come closer to produce the first unit cells of solidification. Numerous alloys exist which contain various percentages of alloying elements and as these alloys cool and change from liquid to solid the atomic structure will alter from a disordered to an ordered arrangement. The degree of interaction will depend upon the compatability of the respective elements.

If cooling curves are drawn for the various alloys, the variation in arrest points at different percentage alloys will produce different types of equilibrium diagrams. Some diagrams show complete insolubility where the elements produce a distinct layered or laminated structure. In contrast, complete solubility provides one single phase comprising of both elements in solution. A combination of both of the above interactions provide a range of alloys which can be mapped out on a partial solubility equilibrium diagram.

When two metals in the molten state are mixed together they may combine totally, i.e. where the atoms of both elements react to form a balanced liquid solution, described as a homogeneous solution. If the elements forming the binary alloy are completely soluble in each other in the liquid state, and if this solubility is maintained during cooling to the solid state, then the equilibrium diagram will appear as shown in figure 3.5.

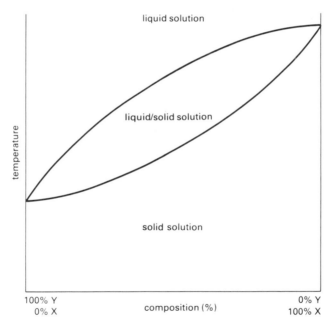

Figure 3.5 Binary equilibrium diagram for metals that are completely soluble in both the liquid and solid states.

This solidified structure does not always result from two metals which combine together in the liquid state. The two metals may be completely insoluble when solid and in such cases the equilibrium diagram is as shown in figure 3.6.

It is also possible for two metals to be partially soluble in one another when solid as shown in figure 3.7. This partial solubility diagram is associated with

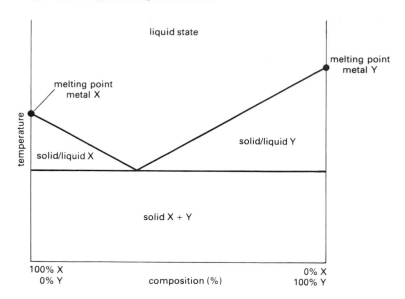

Figure 3.6 Binary equilibrium diagram for metals that are completely soluble in the liquid state but completely insoluble in the solid state.

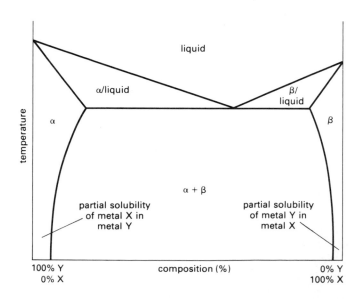

Figure 3.7 Binary equilibrium diagram of partial or limited solubility of metals *x* and *y*.

many engineering alloys, and is considered an asset to the designer, especially where material properties obtained only by heat treatment are desired.

3.6 Complete solid solubility equilibrium diagram

For convenience let us consider an alloy of copper and nickel. These elements combine to produce an alloy, consisting of a single solid solution. The formation of this solid solution is assisted by the fact that both metals possess a FCC atomic arrangement. The foundation structure of the solid solution, the unit cell, is shown in figure 3.8 where some of the nickel atoms have been substituted by copper atoms during solidification. The final structure is described as a substitutional solid solution. Regardless of alloy percentages, the two metals are always completely soluble. As mentioned in Section 3.3 when numerous alloy cooling curves are combined they produce the relevant equilibrium diagram. That for nickel and copper is shown in figure 3.9.

The cooling of the alloy 75% nickel–25% copper, shown in figure 3.9 will now be considered in detail. As the alloy cools from the liquid condition or phase, the liquidus line is cut at temperature T_1 where initial solidification begins. These first dendrites will have a high nickel composition, indicated by S_1 on the diagram. When the temperature has fallen to T_2 the solid now forming is of composition S_2. Whilst the solid being formed is changing in composition along the solidus line, i.e. moving from S_1 to S_2, the remaining liquid surrounding these initial nickel-rich dendrites must also be changing in composition by moving along the liquidus line, thus increasing in copper content. At temperature T_3 the solid forming is of composition S_3 whilst the liquid is of L_2. When temperature T_4 has been reached the last liquid of composition L_3 solidifies and alloy solidification is complete. The overall composition of the alloy is 75% nickel, 25% copper, but it contains dendrites of various composition the core of which will be nickel rich

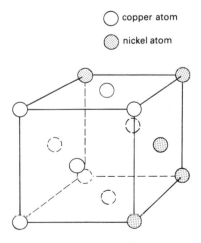

Figure 3.8 Unit cell of copper and nickel atoms forming a solid solution.

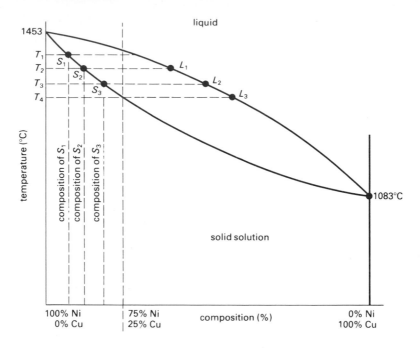

Figure 3.9 Binary equilibrium diagram for nickel and copper.

whilst the outside is copper rich. If the cooling rate is slow enough then atomic diffusion occurs which will result in a balanced atomic dispersion throughout the solid.

3.7   Partial solid solubility equilibrium diagram

Let us consider an alloy of lead and tin where the equilibrium diagram (figure 3.10) produced shows only partial solubility of the metals. The diagram combines features from both the incomplete (cadmium–bismuth, figure 3.3) and the complete (nickel–copper, figure 3.9) solubility diagrams. The cooling pattern of alloy A_1 of figure 3.10 deserves detailed consideration because liquidus–solidus and solvus reaction points are involved. At the liquidus point X dendrites of α solid solution will begin to form; it is important to realise that this newly formed solid is not a pure material but will possess unit cells that contain both atoms of lead and tin. As the temperature continues to fall so the α solid solution increases in size and the amount of liquid becomes less in quantity but richer in tin concentration. With continued heat loss point Y on the solidus line is reached and this is where all the remaining liquid solidifies. The alloy is now completely solid but still quite hot. Further falls in temperature cause no structural changes until point Z on the solvus line has been crossed by the cooling alloy. This point controls the amount of tin that can be absorbed in the α solid solution and as the temperature falls the amount of tin held in solution with the lead also falls.

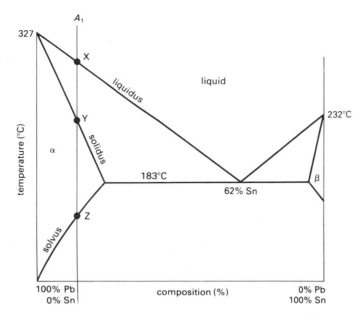

Figure 3.10 Equilibrium diagram for partial solubility between lead and tin.

During the process of pushing out the now surplus tin some lead atoms become entangled, the combined material precipitated out is very rich in tin but contains small quantities of lead and is described as β solid solution, since by definition α solid solution is very rich in lead with small quantities of tin (see figure 3.11). This forcing out or precipitation of β solid solution occurs at the grain boundaries or along certain planes within the α solid solution.

The reaction within the α solid solution at the solvus line, described as *precipitation*, occurs in many other alloys, two of which are aluminium–copper

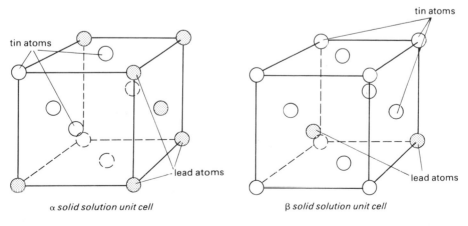

α solid solution unit cell   *β solid solution unit cell*

Figure 3.11 Representations of α and β solid solutions.

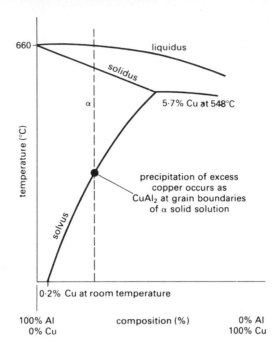

Figure 3.12 Part of the equilibrium diagram for aluminium and copper.

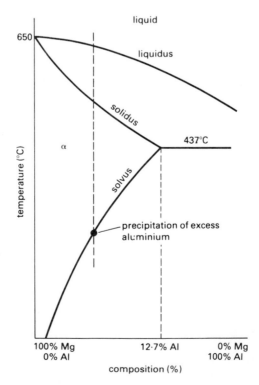

Figure 3.13 Part of the equilibrium diagram for magnesium and aluminium.

and magnesium–aluminium. (See figures 3.12 and 3.13 respectively.) It must be realised that precipitation requires equilibrium cooling conditions. Such a cooling rate is required in order that the atoms can diffuse or migrate through the solid solution. If rapid cooling occurs then precipitation will not occur. This principle of precipitation and the associated properties are dealt with in Chapter 4.

3.8   Purpose of binary equilibrium diagrams

By careful monitoring of the cooling velocities of binary alloys, the temperatures at which specific reactions occur can be assessed quite accurately. Using the cadmium–bismuth equilibrium diagram (figure 3.3), the melting temperature of each major element can be determined, i.e. 100% cadmium at 321°C, 100% bismuth at 271°C. The temperature and the percentage of the elements that will produce a eutectic reaction can also be determined. However, in order to produce such a diagram accurately, equilibrium cooling is essential. Unfortunately such cooling rates are not associated with production foundry casting methods, and are far too expensive, if at all possible. So is the production of such curves a purely academic exercise? If this is the case, what practical purpose do such equilibrium diagrams serve? Their usefulness is in both casting and heat-treatment practice.

Chapter 4

Coring and precipitation processes

4.1   Influence of industrial cooling rates

The previous chapter considered equilibrium cooling conditions which enable atoms to diffuse or move to stable positions within the metal. These theoretical cooling rates allow the atoms to produce a very ordered, balanced arrangement between the major alloying elements. In order to achieve such an internal atomic arrangement the rate of cooling or heat loss has to be extremely slow, thus enabling atomic diffusion or migration. Industrial production methods for casting involve either die or permanent mould casting, sand casting or investment casting (lost wax). In all of these processes there is a variation in the cooling or heat loss rate across the section of the component. The result of these differing cooling rates is that some areas are solid whilst others are still liquid. The crystals produced in one section will therefore vary in composition to the crystals formed in other slower cooling regions, such crystals being said to be 'cored'.

When analysis is required of the composition and properties of untreated alloy specimens, the relevant equilibrium diagram can provide useful information. If a sample is supposed to be 0.9% carbon then microscopic examination should reveal certain structural features according to the relevant equilibrium diagram. The presence of structures known as cementite and pearlite should be visible. (See Chapter 5.) If they do not exist then the specimen coding is incorrect for that sample. The knowledge of the presence of *cementite* and *pearlite* could then be a basis for successful heat treatment and the development of desirable properties of hardness and strength. It is worth noting that the structure produced by rapid quenching 0.9% carbon steel is called *martensite*, which is not found on the equilibrium diagram associated with iron–carbon alloys, because rapid cooling and not equilibrium cooling was used.

Thermal equilibrium diagrams used in conjunction with microscopic examinations can indicate the *working characteristics* of certain metals and alloys. Problems of *hot shortness*, where the metal becomes brittle at elevated temperatures, or *cold shortness*, in which brittleness occurs at lower temperatures can be related to specific areas and transition temperatures on the appropriate equilibrium diagram. Such knowledge can enable preventive heat treatment or alloying procedures to be employed.

4.2   Definition of coring

The term core is associated with the innermost structure of an object. We refer to the Earth's core, an apple core and being able to appreciate the core of a complex problem. These examples conjure up pictures of the internal arrangement, where development or growth began. If we consider the formation of crystals during solidifcation, where equilibrium cooling does not exist, then a structure will be produced where the composition across the crystals will vary. Such crystals will possess a cored structure. The first material to solidify will be the element with the highest melting point, thus the nuclei of the dendrites will have a high concentration of one of the alloy's elements, the net result is that the last material to solidify must be rich in the second element. This variation in structure, called

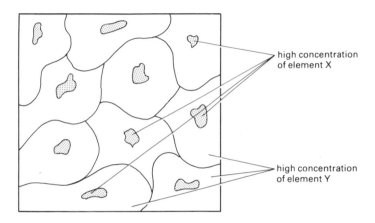

Figure 4.1 Unequal distribution of elements causing coring.

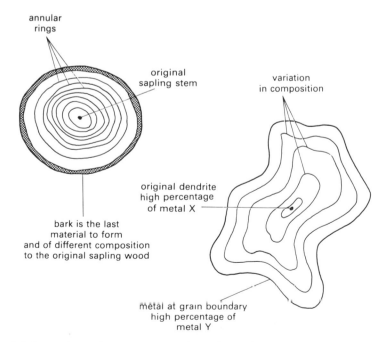

Figure 4.2 Comparison of tree structure and cored crystal.

coring, is due to the lack of atomic diffusion, resulting from non-equilibrium cooling rates. Such a cored arrangement is shown in figure 4.1. The cross sectional structure of a tree trunk may be likened to the cored crystal: on the outside is the bark which is different in composition and appearance to the innermost section of the trunk and the annual rings show changing composition across the trunk section. The original sapling formed the foundation of the tree and may be considered to have the same function as the initial dendrites, i.e. a foundation upon which to develop and grow (see figure 4.2).

4.3   Mechanism of coring

Coring arises because cooling rates are too fast to allow atomic diffusion to occur during solidification. The structural build-up of a crystal can best be explained by referring to figure 4.3. When the alloy of composition C_x starts to cool, the first dendrites are formed at temperature T_1, where the cooling liquid crosses the liquidus line at L_1. The composition of these initial dendritic particles can be obtained by projecting across from L_1 on the liquidus line, to position S_1 on the solidus line and then down vertically to position C_1 on the composition axis. At

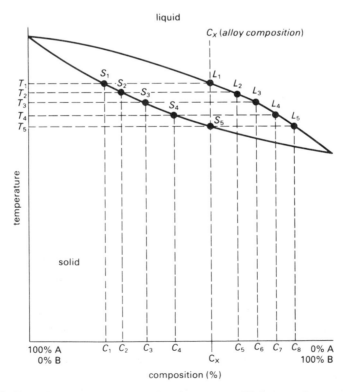

Figure 4.3 Binary equilibrium diagram showing non-equilibrium cooling rates that will produce a cored structure.

temperature T_2 the solid, S_2, is of composition C_2 while the liquid composition had been changed from C_x to C_5. At temperature T_4 the solid material being deposited out of solution is of composition C_4 and the remaining liquid, which by now is quite small in comparison to the solid is of composition C_7. During the cooling process the liquid becomes progressively richer in element B so that when solidification temperature T_5 is reached the last drop of liquid will be of composition C_8. A high percentage of element B will be in this last liquid. The cored structure outlined above has initial grain nuclei rich in element A while the last solid to form at the grain boundaries is rich in element B. Although a concentration gradient exists across the grain's structure, the overall or average composition will be C_x. The composition line C_x can be considered as a fulcrum of a balance where the composition of the solid formed is balanced by the composition of the remaining liquid until solidification is complete (see figure 4.4).

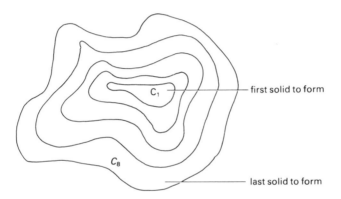

Figure 4.4 Typical cored structure showing composition variation from C_1 rich in metal A to C_8 rich in metal B.

4.4   Elimination of coring

When a cored structure is produced as a consequence of cooling that is not in equilibrium (figures 4.5 and 4.6) the variation in the composition of the elements can have serious effects upon the mechanical and physical properties. The strength of the alloy, its corrosion resistance and its stability at elevated temperatures can all be impaired. Because cored crystals are composed of areas rich in one element, the strength of such crystals will vary across their section. As a result the overall strength of the cast component will be affected. The high concentration areas can produce *anodic* and *cathodic* regions within the crystals and eventually corrosion breakdown can occur in the component. It is the alloys which possess a wide solidification or freezing range which are prone to *element segregation*.

Copper–tin alloys have a wide melting point difference and when poured into a mould cavity the element with the higher melting point will solidify first—the

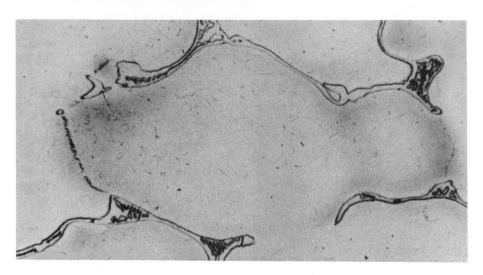

Figure 4.5 Cored structure showing uneven distribution of elements. Magnification × 500.

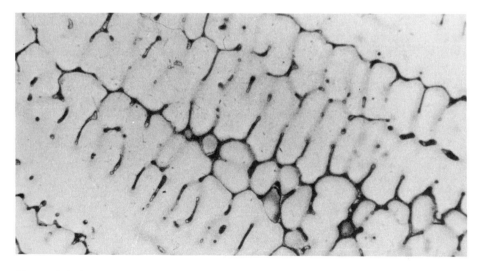

Figure 4.6 Coring within a solid solution due to non-equilibrium cooling rates. Magnification × 100.

copper. In consequence the lower-melting tin will become concentrated in the inner central regions of the cast component. It is possible for segregation or coring mechanisms to affect individual crystals.

As shown in Chapter 2 during dendritic solidification the element with the lowest melting temperature will solidify *last*—at the crystal boundaries. If a component comprised of such a structure is heated, then at a certain temperature the grain boundary areas will become molten, producing a ribbon of molten metal around each solid crystal. Such a condition provides little plasticity at elevated temperatures and is described as *hot short*. It is therefore necessary to eliminate such cored structures.

Due to coring it is impossible to cold work certain cast alloys, one typical alloy being copper–tin, which produces a tin bronze. The coring mechanism produces brittle copper–tin compounds which are segregated between the crystals of the alloy. The presence of such brittle constituents causes *cold shortness*, i.e. the ingot cracks during cold rolling.

An *homogenising anneal* is applied in order to eliminate or reduce coring by inducing atomic diffusion, which leads to uniformity of composition throughout the casting. The time and temperature used depends upon the rate of diffusion of the *solute* atoms in the crystals of the *solvent* metal. In alloys of copper and zinc, the zinc diffuses readily through the copper, whilst with copper and tin, the diffusion of migration of the tin is very slow. Copper–tin alloys (bronzes) are given an homogenising anneal at temperatures of 650–800°C for up to 7 h. This process can also be used on alloys of nickel–silver and copper–nickel.

With steel alloys the casting must be heated to a temperature above the upper critical point i.e. a temperature which allows atomic diffusion to occur thus developing an even distribution of the atoms throughout the structure of the alloy, and maintained at that temperature for a sufficient period of time before cooling in still air. The casting should be maintained at the homogenising temperature for at least 1 h per 25 mm of thickness where temperatures of 1000°C and higher are used.

4.5 Definition of precipitation

An important feature of the partial solid solubility diagrams discussed in Chapter 3 were the solvus lines. These lines indicate the degree of solubility or absorption of one element within the lattice arrangement of another. The type of structure so produced is described as a solid solution. The first solid solution to form is described as α, the second as β and so on. The slope of the solvus line indicates that the solubility level varies with temperature. As the temperature increases the degree of solubility inceases. When the rate of cooling is slow, the 'excess' element is forced out of the solid solution as the temperature falls. This mechanism is called *precipitation*.

Precipitation can be clearly observed when salt and water are combined. At room temperature a certain amount of salt can be dissolved in a beaker of cold water. If extra salt is added, the crystals remain visible in the solution. If the water is heated the excess crystals disappear into solution, only to reappear by precipitation when the temperature falls. This is illustrated in figure 4.7. At temperature T_1 an amount of salt, S_1, can dissolve in the water. At temperature T_2 the concentration increases to a value S_2. The change in the absorption rate is reversed when the temperature falls and so precipitation occurs. The concentration values S_1 and S_2 are connected by the curve called the solvus line.

A reaction similar to the salt and water model can occur in alloys that contain solid solutions or ones in which the rate of cooling is very slow. It is obvious that precipitation in metals will be very slow compared with the precipitation in

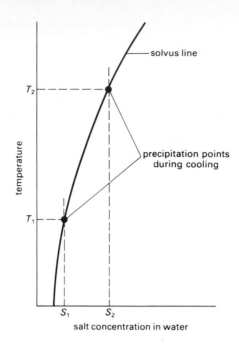

Figure 4.7 Temperature change and solubility levels.

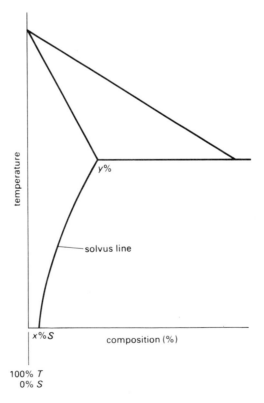

Figure 4.8 Typical equilibrium diagram showing changing solubility with changing temperature. The change in solubility is indicated by the change in the slope of the solvus
line.

saturated salt solutions. It is possible to retain a saturated solid solution by rapid quenching where the resulting structure may develop desirable mechanical properties e.g. increased hardness and strength. This type of saturated solid solution will only result if the amount of alloying element is in excess of x% as shown in figure 4.8. The value of x will depend upon the particular alloy. The precipitation mechanism involves the cooling rate and solubility levels of the alloying elements.

4.6   Mechanism of precipitation hardening

This process is associated with alloys which are capable of forming partial solid solubility. By heat treating the alloy the final structure is harder than that obtained by simple cooling.

In figure 4.9 the solubility of metal Y in metal F at 0°C is shown as 0.2% Y rising to 5.7% Y as the temperature increases. This changing solubility of one element in another is the hinge pin of the process. If an alloy containing 4.5% Y in this example is heated to 500°C, the element will become integrated with the

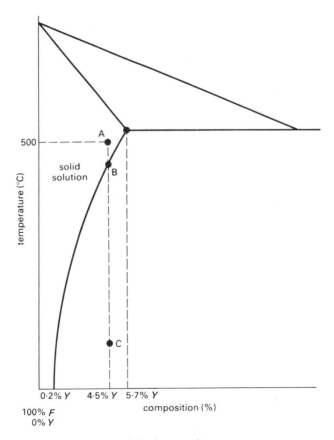

Figure 4.9 Effects of cooling on precipitation reaction.

metal *F* forming a solid solution, i.e. atoms of element *Y* form part of the lattice structure of element *F*. If the alloy cools very slowly, a point B on the solvus line will eventually be reached indicating the solubility limit of saturation point of element *Y* in the element *F*. At this point precipitation would begin. However, if precipitation is not allowed, i.e. the alloy is rapidly cooled from 500°C by quenching in water, then the 4.5% *Y* is kept in solid solution. Therefore the content of element *Y* at room temperature has been dramatically increased from around 0.2% to 4.5%. The lattice structures are severly strained and this condition effectively increases the hardness of the alloy. Typical structures are shown in figure 4.10 which correspond to positions A, B, and C in figure 4.9.

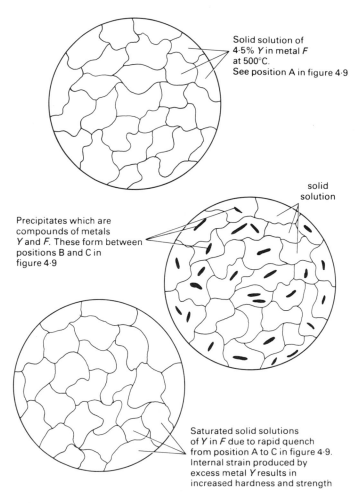

Solid solution of
4·5% *Y* in metal *F*
at 500°C.
See position A in figure 4·9

solid
solution

Precipitates which are
compounds of metals
Y and *F*. These form between
positions B and C in
figure 4·9

Saturated solid solutions
of *Y* in *F* due to rapid quench
from position A to C in figure 4·9.
Internal strain produced by
excess metal *Y* results in
increased hardness and strength

Figure 4.10 Effects of temperature and cooling rate on structure.

4.7   Advantages and limitations of precipitation

Preventing precipitation by rapid cooling enables metals such as aluminium alloys to be hardened. Without this reaction such materials would remain soft, thus limiting their range of engineering applications. Whilst increasing the hardness, the strength of the aluminium alloy is also improved (discussed in detail in Chapter 8). If precipitation does occur the resulting structure could be relatively soft. This lack of strength is due to the precipitated particles. In some alloys, such as aluminium–copper alloy, the precipitate depletes the solid solution of copper leaving behind a very rich aluminium matrix. The combined effect of this is the formation of a weak material.

Whilst the prevention of precipitation improves hardness and strength qualities, components made from these alloys, especially aluminium–magnesium–zinc alloy, must be formed into shape as soon as possible after solution treatment, i.e. heating to obtain a solid solution followed by a rapid quench. Problems of storage of the unformed alloy can cause a serious delay in production processes. The development of refrigerated storage for aluminium rivets in the aircraft industry goes towards solving this problem.

Chapter 5

Iron–carbon alloys

5.1   Composition and structure

Pure iron is a soft metal possessing a body-centred atomic arrangement (BCC) (see Section 2.3) and in this condition it is of little use to the engineer. As such, it is described as α iron (alpha) or ferrite and is considered to possess no carbon. Iron has the ability to change its atomic arrangement while heat energy is being applied. This ability, called *allotropy*, enables iron–carbon alloys to be readily heat treated. At room temperature the iron is BCC and this structure is maintained until a temperature of 910°C is reached whereupon the atoms rearrange themselves to form a face-centred cubic arrangement (FCC).

Because 14 atoms are required to form a dense FCC unit cell and only nine are required to form an open BCC cell there is a reduction in volume as shown in figure 5.1.

The FCC arrangement is called *austenite* and is indicated by the symbol γ (gamma). In this condition the material can absorb into solution up to 1.7% carbon. This maximum solubility corresponds to a temperature of 1147°C.

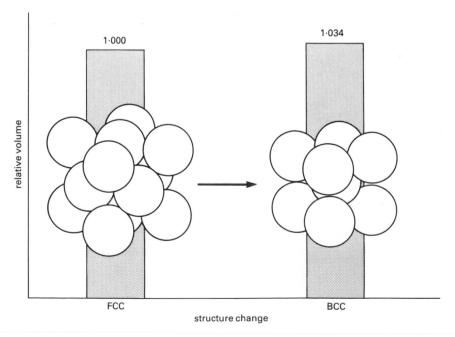

Figure 5.1 Change in volume with change in unit cell formation.

As discussed in Section 4.7, the solubility value changes with temperature along the solvus line. These two lattice formations of BCC and FCC that exist between room temperature and 1000°C are of great importance in the development of particular mechanical properties. The ferrite lattice changes back to BCC at 1390°C. The structure is then called δ iron (delta).

One other change in the character of iron occurs at 768°C when it loses its magnetism but no crystal change or lattice reordering occurs. Therefore mechanical properties are not affected and the iron in this condition is known as β iron (beta).

The dimensional changes due to lattice reformation and subsequent volume changes can be detected using an instrument called a *dilatometer* (figure 5.2).

If readings of dimensional change against temperature are recorded and plotted, the resulting curve shows clearly the volume changes due to atomic reorganisation (figure 5.3a and b).

The carbon atom is small in comparison to the atoms of ferrite and can therefore position itself between the ferrite atoms. Since the space between the atoms is called the interstices then this carbon–ferrite formation is described as an intersitial solid solution (figure 5.4). As the levels of carbon content are increased the melting temperature and the upper critical temperature of the iron–carbon alloy decreases (figure 5.5 and see Chapter 3).

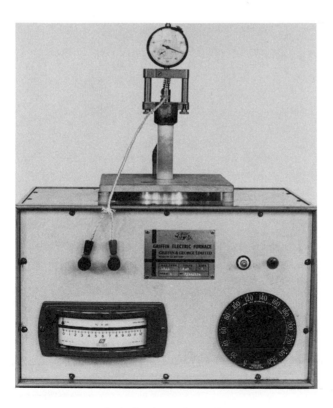

Figure 5.2 (a) Dilatometer used for indicating volume change due to atomic rearrangement.

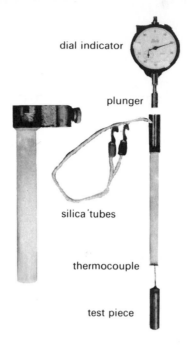

dial indicator

plunger

silica tubes

thermocouple

test piece

Figure 5.2 (b) Important component parts of a dilatometer.

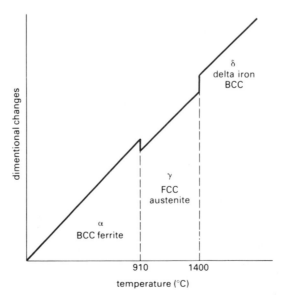

δ
delta iron
BCC

dimentional changes

γ
FCC
austenite

α
BCC ferrite

910   1400

temperature (°C)

Figure 5.3(a) Change of unit cell with increasing temperature.

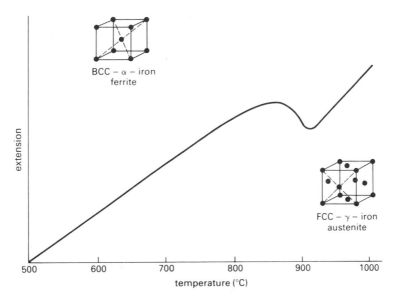

Figure 5.3(b) Volume change due to lattice reorganisation. Extension versus temperature for 0.15% carbon steel using dilatometer to indicate allotropic change.

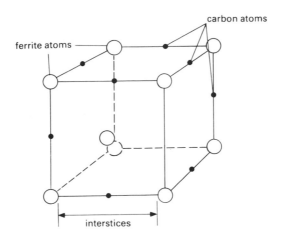

Figure 5.4 Body-centred cubic iron containing 0.02% carbon in solid solution. α Iron, an interstitial solid solution.

The varying concentration of carbon produces different classes of iron–carbon alloys. With a content of up to 0.05% carbon and the material is classified as pure iron; between 0.05% and 2% carbon the alloy is classed as a steel; between 2% and 4.3% carbon the material is classified as cast iron. The actual reaction that occurs between the carbon and the ferrite produces a compound, metallic in nature known as iron carbide or cementite, chemical formula Fe_3C. The amount of carbon will control the amount of cementite in the alloy and because the

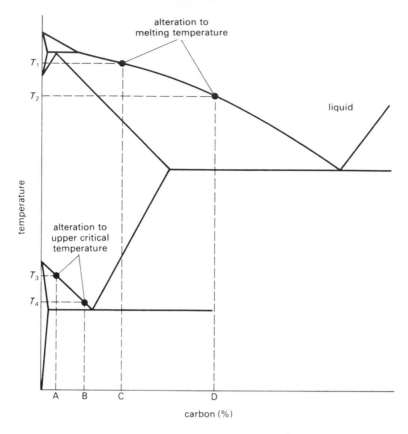

Figure 5.5 Effect on melting temperature and upper critical temperature with varying carbon content.

cementite is hard and brittle the mechanical properties will change according to carbon content. Consider in detail three alloys of 0.2% carbon, 0.83% carbon and 1.2% carbon cooling from 1100°C. The associated equilibrium diagram is shown in figure 5.6.

At a temperature of 1100°C all three alloys are in the form of gamma iron (γ) or austenite which is a FCC solid solution of carbon in iron (figure 5.7). The lowest temperature for an alloy of iron and carbon to exist in the form of 100% austenite or γ iron is 723°C and the alloy composition is ferrite containing 0.83% carbon as shown in figure 5.6.

At 723°C the austenite changes to a structure known as pearlite, this description is given because of the surface lustre which resembles 'mother-of-pearl'. The structure of pearlite consists of alternate layers of ferrite and cementite. The reaction that produces this pearlite is described as a *eutectoid reaction*. The structure will be 100% pearlite (figure 5.8a and b).

All alloys cooling from the austenitic range down to room temperature are 'compelled' to produce some pearlite as part of their final structure. If we consider the alloy containing 0.2% carbon we quickly realise that the initial carbon

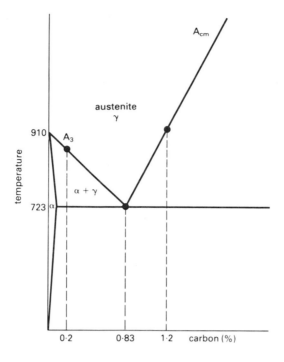

Figure 5.6 Part of the carbon equilibrium diagram showing the cooling of three alloys of compositions 0.2, 0.83 and 1.2% carbon.

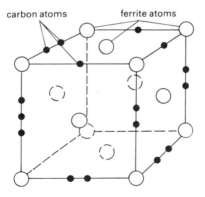

Figure 5.7 Face-centered solid solution of carbon in ferrite. Described as gamma iron (γ) or austenite.

content of the alloy is below the magical 0.83% carbon value. During cooling from the austenitic condition the 0.2% 'carbon' alloy precipitates or 'forces' ferrite out of solid solution, and the remaining austenite gradually has its carbon content increased. Whilst this ferrite is precipitating the austenite gets smaller in size, although its percentage carbon is increasing. Eventually small particles of

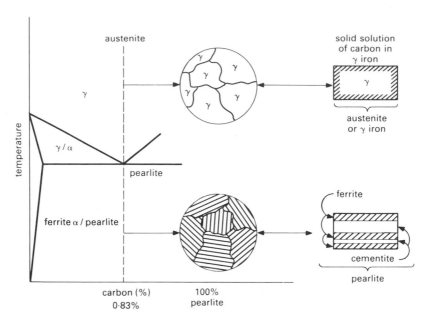

Figure 5.8(a) Cooling curve for 0.83% carbon steel.

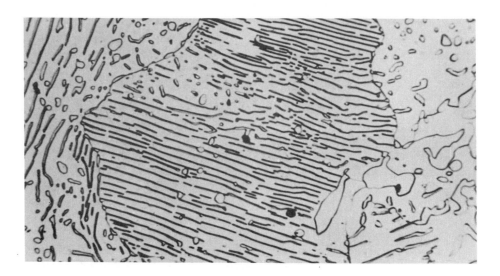

Figure 5.8(b) Eutectoid structure 0.83% carbon. Magnification × 1000.

austenite exist containing 0.83% carbon which then change to particles of pearlite. This pearlite or eutectoid reaction occurs at 723°C. The final structure will be large quantities of ferrite and very small quantities of pearlite as shown in figure 5.9.

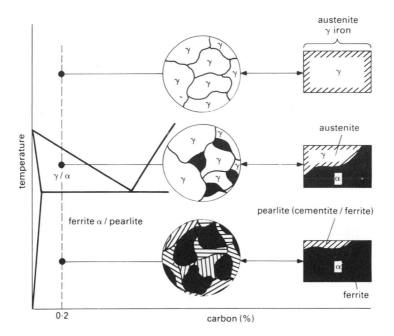

Figure 5.9 Cooling of 0.2% carbon alloy.

An alloy that contains less than 0.83% carbon is described as a *hypoeutectoid*. With the exception of pure iron no matter how small the carbon content in the alloy, there will always be some pearlite present in the final structure, providing the rate of cooling allows atomic diffusion to occur.

With the 1.2% carbon alloy this contains in excess of 0.83% carbon and therefore during cooling from the austenitic range this extra carbon must be removed from the austenite. The excess carbon is precipitated out of solid solution but during the process it brings with it ferrite atoms and the final precipitate is not pure carbon but is cementite (Fe_3C). As cooling and cementite precipitation continue the remaining austenite becomes gradually weaker in carbon content until only 0.83% carbon exists. Then it changes to pearlite the reaction temperature being 723°C. The final structure is cementite and pearlite (figure 5.10a and b). When the initial carbon content is above 0.83% carbon such an alloy is described as *hypereutectoid*.

The precipitation products, i.e. ferrite from 0.2% carbon alloy and cementite from 1.4% carbon alloy occur around the grain boundaries of the austenite and the final structure will be a network of precipitates around the grain boundaries of pearlite. Such a condition will effect the mechanical properties of the alloys.

5.2 Structure and properties

The nature and characteristics of all materials depend totally upon the atomic arrangements within the structure of the material. In the iron–carbon alloy range,

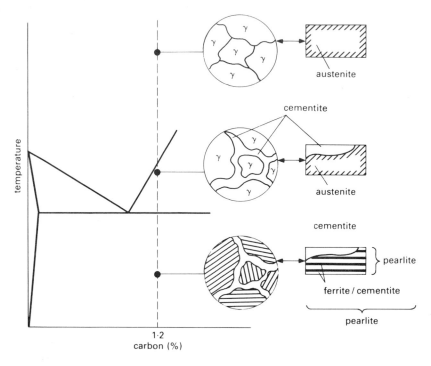

Figure 5.10(a) Cooling curve for 1.2% carbon alloy.

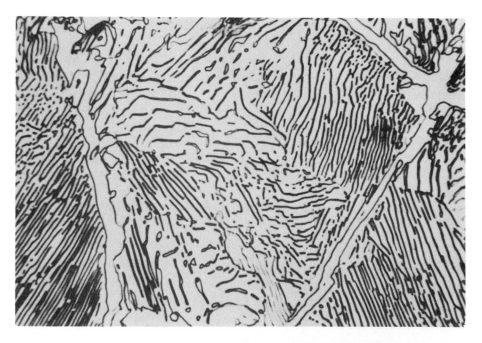

Figure 5.10(b) 1.2% carbon steel cooled slowly from 980°C showing cementite (Fe_3C) precipitate at grain boundaries of pearlite. Magnification × 1000.

Carbon content (%)	Classification	Application
0.05–0.15	Dead mild steel	nails
0.1–0.2	Mild steel	RSJ sections
0.2–0.3		Gears, forgings
0.3–0.4	Medium carbon	Axles, connecting rods
0.4–0.5		Crankshafts
0.5–0.6		Wire ropes
0.6–0.7	High carbon	Set screws
0.7–0.8		Hammers
0.8–0.9		Cold chisels
0.9–1.0	Tool steels	Axes
1.0–1.1		Drills and taps
1.1–1.2		Ball bearings, lathe tools
1.2–1.3		Files, reamers
1.3–1.4		Razors, saws

Figure 5.10c Applications of various iron-carbon content alloys

carbon atoms are introduced to the ferritic or iron matrix, these carbon atoms interact with the ferrite atoms and such interactions together with cooling rates are responsible for the range of properties associated with such alloys, i.e. hardness and strength. When the carbon content is below 0.15% the material is described as 'dead mild steel'. The tensile strength of this alloy, 280 N/mm^2, is low in comparison to the higher carbon-content alloys but it does possess relatively good ductility. An application of this alloy would be the nails required for the wood-working industry.

By increasing the carbon content up to 0.3% the structure becomes stronger. This alloy is known as mild steel and is used for forging and in the manufacture of gears etc. Steels containing less than 0.3% carbon do not quench-harden satisfactorily. By increasing the carbon content up to 0.6% a range of alloys known as medium carbon steels are produced. Industrial applications are axles and wire ropes (see figure 5.10c). If these alloys were to be examined microscopically the existence of a lamination-type structure would be observed (figure 5.11) the presence and quantity of this layered structure of cementite and ferrite is responsible for the degree of strength and hardness (tensile strength 700 N/mm^2). It is important to realise that as the carbon content of an iron–carbon alloy increases it combines with the ferrite to reform at room temperature as pearlite or cementite or as a combination of both constituents. The carbon does *not* exist as a separate entity as shown in figure 5.12.

Ductility or percentage elongation falls as carbon content increases, this is because atomic slip which is responsible for ductility becomes more restricted with higher carbon content. The 'restriction' to slip is also responsible for increasing the hardness and tensile strength (figure 5.13).

When alloys containing more than 0.83% carbon are subjected to rapid cooling rates a structure can be developed that possesses the ability to function as a cutting tool.

Figure 5.11 Existence of layered structure pearlite (cementite/ferrite) with ferrite. Magnification × 250.

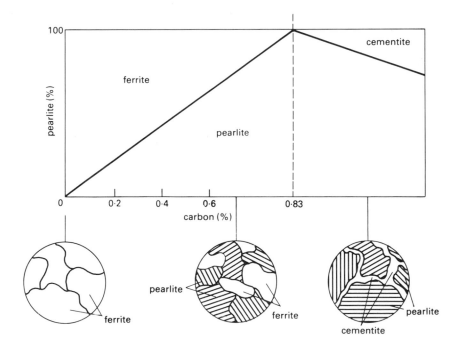

Figure 5.12 Structural changes due to carbon content variation.

Let us recap on the major constituents of all iron–carbon alloys, i.e. ferrite and cementite. If the alloy contained cementite only it would be extremely brittle, with ferrite only it would be extremely soft. When combined to form pearlite the desirable properties of each are combined. If cooling rates not associated with the

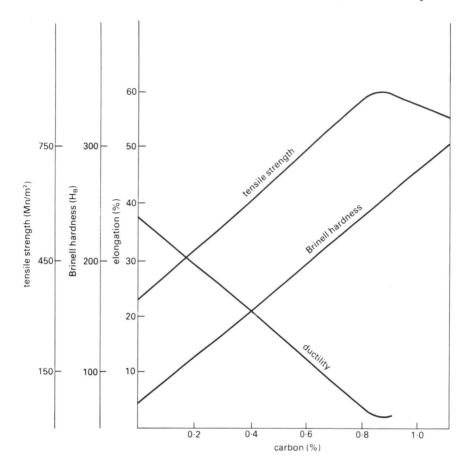

Figure 5.13 Effect on mechanical properties with varying carbon content.

equilibrium conditions shown in figures 5.8 to 5.10 are used then special properties which produce varying degrees of hardness are obtained.

The selection of an iron–carbon alloy will depend upon its application. For ships plate known as 'boiler plate' then a steel with 0.2% carbon would be satisfactory. Such an alloy would allow quick cheap fabrication to be carried out. These manufacturing qualities are essential in the ship-building industry. Steel with 0.4% carbon is considered twice as strong as pure iron, while 1.0% carbon is three times as strong, but with increasing strength there is a reduction in ductility. Such carbon contents should be associated with components that are not required to be ductile, i.e. 'fine edged' cutting tools, such as razors, where the cutting edge must be maintained.

5.3 Iron–carbon equilibrium diagram

Steels are basically alloys of iron and carbon up to 1.5% carbon. It is convenient, therefore, to consider only a portion of the iron–iron carbide (Fe–Fe$_3$C) diagram

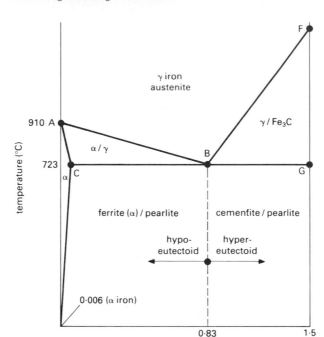

Figure 5.14 Equilibrium diagram for iron and carbon showing steel portion.

(the complete diagram is discussed in detail in Section 7.2). The 'steel' section of the diagram is also of interest when discussing heat treatment. The diagram area will therefore be restricted to 1.5% carbon and 1000°C (figure 5.14).

From figure 5.14 it will be seen that α iron or ferrite can only hold 0.006% carbon in solid solution at room temperature. This increases to 0.04% carbon at 723°C, this structure alone has little practical engineering applications. However, gamma iron γ or austenite can hold in solution considerably more carbon and increases the scope of engineering applications. At a temperature of 723°C the carbon content of austenite is 0.83% rising to 1.7% at a temperature of 1130°C. Alloys below the eutectoid carbon content are described as hypoeutectoid whilst alloys containing more that 0.83% carbon are termed hypereutectoid. By increasing the carbon content, the transition from α iron to γ iron is reduced. This is indicated by the line AB in figure 5.14. This line is also described as the upper critical point or the A3 line. Above this line the iron–carbon alloy is completely austenite. At point B the eutectoid reaction produces a structure from austenite described as pearlite. This reaction occurs on the lower critical point line BC which is termed the A1 line, below the A1 line on the iron–carbon diagram austenite does not exist; it becomes pearlite. Between the A3 and the A1 lines the material becomes non-magnetic, this occurs at 769°C and is described as the Currie temperature which is denoted as the A2 point. At the A2 point this loss of magnetism is not as important to the engineer as the critical points of A1 and A3. Below the A1 or lower critical point line the structure is ferrite (iron) with pearlite.

For hypereutectoid alloys, i.e. containing more than 0.83% carbon, the line BG is considered as the lower critical point whilst the line BF is the upper critical point or the A$_{cm}$. This upper critical line BF may be considered as a solvus line which monitors the saturation level of carbon in austenite. When cooling passes through BF the austenite precipitates carbon in the form of Fe$_3$C. Whilst cooling continues for these carbon-rich alloys carbon is being continually precipitated as Fe$_3$C which has the effect of reducing the carbon content of the remaining austenite to that required to form pearlite, i.e. 0.83% carbon. This reaction, termed a eutectoid reaction, takes place at 723°C when the remaining austenite transfers to pearlite. The structure at room temperature for such an alloy would be Fe$_3$C together with pearlite.

The iron–carbon diagram can be obtained by measuring the arrest points during heating or cooling and comparing the results. If the specimens are monitored during heating then these arrest points are described as A$_{c1}$, A$_{c2}$ and A$_{c3}$. If the arrest points were obtained during cooling then they would be shown as A$_{r1}$, A$_{r2}$, A$_{r3}$. Concerning the above arrest symbols the A denotes arrest (from French arrière) the symbols c from chauffage which is heating and the symbol r from refroidissment, which means cooling.

5.4   Definitions of constituents

Ferrite

Body-centred at room temperature and can only absorb extremely small quantities of carbon in solution. It is allotropic, changing to FCC at 910°C and back to BCC at 1400°C. At room temperature it is quite ductile but not very strong. It is indicated by the symbol α, and does not exist above 910°C.

Austenite

The structure is formed by equilibrium cooling rates. It possesses a face-centered unit cell and is an interstitial solid solution of carbon in a ferrite matrix. Formation of this structure commences at 723°C, upon cooling below this temperature it forms as a pearlitic structure. The material is non-magnetic. Due to its increased carbon abosrption level the austenite structure is extremely important in the development of hardened structures.

Cementite

This is a hard, brittle compound with chemical formula Fe$_3$C. It is created during cooling austenite where the solubility of carbon in ferrite has been exceeded. For its formation it requires at least 0.83% carbon.

Pearlite

Formed at the lower critical temperature of 723°C from austenite of carbon content 0.83%. The reaction to form pearlite is described as the eutectoid reaction and produces a laminated structure. This structure contains alternative layers of α iron adjacent to the iron carbide compound Fe_3C.

5.5   Iron–carbon alloys and heat treatment

When very slow cooling rates are applied to iron–carbon alloys, the austenite that exists above the upper critical point, changes its structure. The final structure could contain varying quantities of ferrite, pearlite and cementite, the formation of these structures being dependant upon carbon content. The reason such changes can form is because the very slow cooling rate enables atomic diffusion to take place. When cooling, the transition points are described as A_{r3} A_{r1} and A_{rcm}, when heating the transition points are A_{c1} A_{c3} and A_{ccm} (figure 5.15).

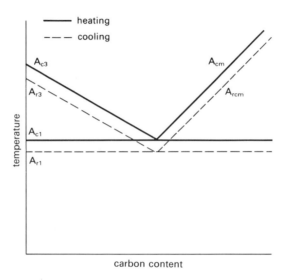

Figure 5.15 Arrest points for heating and cooling of iron carbon alloys.

The equilibrium diagram produced from noting the arrest points of the iron–carbon alloys can only 'predict' structures that are obtained under similar slow-cooling conditions. Industrial heat-treatment processes that come close to equilibrium cooling are the annealing techniques (discussed in section 5.6). Structures arising from such processes may be 'mapped out' on the equilibrium diagram. Those that are rapidly quenched cannot be described from an equilibrium diagram because atomic diffusion has not been allowed to occur. Such accelerated cooling rates will produce different microstructures possessing different properties. One typical, extremely hard, microstructure is called martensite and is produced by rapid cooling from the austenitic condition. If slight modifications to the cooling rate are made then corresponding changes in the degree of hardness will result. This development of structural change is extremely important to the engineering industry.

The S-curve described because of its shape, is also known as the temperature transformation curve. These curves are associated with isothermal temperatures i.e. where the temperature is constant across the section. Such curves are just as important for the heat treatment of steel as the iron–carbon equilibrium diagram. This curve predicts the type of structure possible due to changing the rate of cooling see figure 5.16. Each iron–carbon alloy has its own S-curve therefore *all* the structures resulting from differing cooling rates or velocities can be predicted.

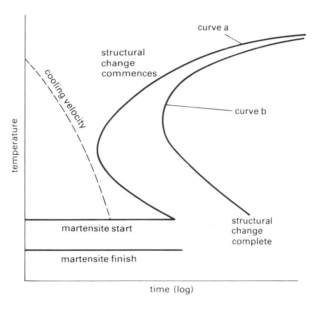

Figure 5.16 Typical S curve.

By combining the equilibrium diagram and the appropriate S-curve, all the important factors concerning the types of heat treatment processes for a particular iron–carbon alloy can be shown, i.e. transition temperatures, rate of cooling and final microstructure.

5.6 Annealing processes

Annealing can be used to give steel soft, ductile properties, thus allowing further cold-working operations to be conducted. The process may also be applied to large castings and ingots in order to obtain a more uniform crystal structure. Annealing is also recommended for stress relief in castings and fabricated structures. The resulting crystal structures differ with different applications and the annealing temperature required must also differ. This variation in operating temperature is also dependant upon carbon content (figure 5.17), i.e. the upper critical point for 0.1% carbon is higher than for 0.4% carbon. The upper critical point is the temperature at which the structure of the iron–carbon alloy changes from ferrite combined with austenite to one of only austenite.

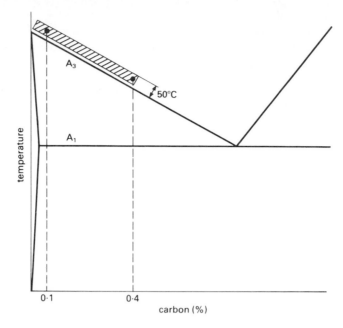

Figure 5.17 Full annealing range for iron–carbon alloys.

In order to standardise the operating temperature with its desired properties and crystal structure, the 'general annealing' process has been subdivided into full, process and spheroidise annealing. Each process will now be considered in detail.

Full annealing

For industrial applications full annealing is only applicable to steels of less than 0.4% carbon, since excess cementite is developed from higher-carbon alloys to the detriment of the final annealed structure. The steel is heated to 50°C above the upper critical point (A_3), this temperature being dependent upon the carbon content (figure 5.17).

Once at this temperature the steel is held or *soaked* to attain uniform temperature throughout its section. The steel is now cooled very slowly during which time atomic diffusion occurs, transforming austenite to ferrite and pearlite. The rate of cooling is as close to equilibrium cooling as is possible, and is achieved by cooling in the furnace, or by 'burying' in ash or sand. The result is a very soft steel, comprising of ferrite, precipitated at the grain boundaries of the original austenite, which breaks down into pearlite. Figure 5.18 illustrates pearlite grains surrounded by ferrite.

Process annealing

This heat treatment process is applied to steels containing less than 0.3% carbon that have been cold worked and are required to be further cold worked. The final

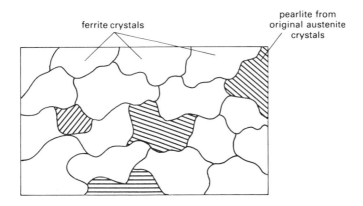

Figure 5.18 Typical structure produced from 0.4% carbon steel after full annealing.

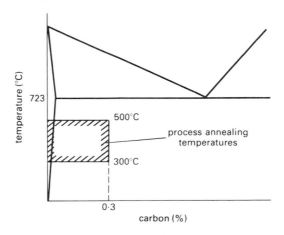

Figure 5.19 Process annealing range for low-carbon alloys.

structure is required to be strong and sufficiently ductile to allow further crystal deformation by working. The treatment involves heating to between 350 and 500°C, this temperature is dependant upon the degree of previous cold working. Because the process temperature range is below the lower critical point A_1 the technique is sometimes described as sub-critical annealing (figure 5.19). Whilst the heat treatment is being conducted slight recrystallisation occurs within the deformed ferrite crystals, this structural change will occur at the *recrystallisation temperature*. If tremendous levels of cold working occurred then the deformed crystals will recrystallise at a low temperature and vice versa. During the recrystallisation of the ferrite grains the heavily deformed pearlite areas are unchanged. The final structure produced by process annealing will be small, recrystallised ferrite grains surrounded by distorted pearlite regions (figure 5.20). The effect of such a structure is to restore a degree of ductility whilst maintaining the strength.

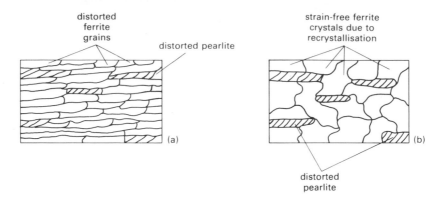

Figure 5.20 Effect on cold-worked structure after process annealing. (a) Typical cold-worked, low-carbon alloy. (b) Structure after process annealing.

Spheroidise annealing

This process is limited to steels in excess of 0.5% carbon and consists of heating the steel to 700°C (figure 5.21) where upon the material is soaked. At this temperature any cold-worked ferrite will recrystallise and the iron carbide present in the pearlite will form as spheroids or 'ball up'. As a result of the recrystallisation of ferrite and the breakdown of the pearlite into spheroids of cementite or carbide, the strength and hardness are reduced. Such a structure develops good shearing qualities which are required in metal-machining techniques. The spheroid form of the carbide within the structure, assists the shearing action of the cutting tool, by providing internal stress concentrations which compliment the cutting action of the tool. The most important application of this process is to improve the machineability of high-carbon steel. Spheroidise annealing requires less time than the full annealing process.

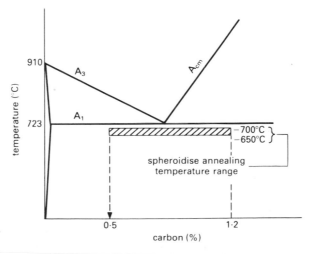

Figure 5.21 Spheroidise annealing position on thermal equilibrium diagram.

5.7 Normalising

This process is primarily associated with low- and medium-carbon steels. The heat-treatment temperature for the normalising process is within the same range as full annealing i.e. within the austenitic range (figure 5.22). The actual temperature varies with carbon content. Once at the required temperature the component is soaked, to allow for a uniform temperature to be obtained. After soaking the component is cooled in *still* air which is a faster cooling rate than for annealing. This shorter cooling time produces a small grain size as there is little time for grain growth to occur. The small-grain structure is responsible for improving the machinability and for providing a slight increase in strength and hardness. The final structure is of a small grain size with a fine pearlite structure. Normalised steels are always harder and stronger than equivalent carbon-content steels in the annealed condition. Although normalising incorparates a faster cooling rate than annealing, the rate of cooling is slow enough to allow atomic diffusion. Therefore, the appropriate section on the iron–carbon equilibrium diagram can be used to predict the type of structure produced by such a cooling rate, i.e. ferrite and pearlite. This structure is associated with the slow cooling of hypoeutectoid steels.

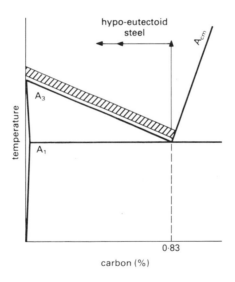

Figure 5.22 Normalising range for low-and medium-carbon steels.

5.8 Quench hardening

This process requires the heating of a high-percentage carbon alloy to above the upper critical point, where the ferrite and pearlite are transformed to austenite. In the austenitic condition the carbon is evenly dispersed amongst the spaces or interstices of the FCC unit cells. As a result of rapid cooling, the austenite cannot transform back to ferrite and pearlite by atomic diffusion and a very hard structure, martensite, which appears to have a 'needle like' structure is produced. Since atomic diffusion is severely restricted by the rapid quench, the carbon atoms become trapped. The high cooling rate also restricts the transition back to a BCC arrangement. This restriction, in conjunction with the carbon atom 'fixing', produces a severely distorted atomic arrangement. Such a distorted atomic structure is unable to 'slip' or deform easily and is responsible for developing the hardness property. For hypoeutectoid steel i.e. less than 0.83% carbon the heating temperature is 50°C above upper critical point, for hypereutectoid the temperature is 50°C above the lower critical point (figure 5.23).

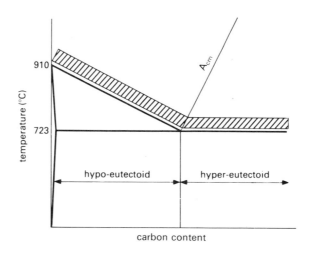

Figure 5.23 Hardening range for carbon steels.

5.9   Critical cooling rate

In all heat-treatment processes, the final structure, and hence the desired properties, are governed by the rate of cooling from a preselected temperature (figures 5.17, 5.19, 5.21 to 5.23). With the quench-hardening process the speed of quenching can affect the amount of martensite formed. This severe cooling rate will be affected by the component size and the quenching medium which may be brine, water, oil or air. The critical cooling rate or velocity is the slowest speed of quenching that will ensure maximum hardness. This critical cooling velocity is shown on the 'S' curve in figure 5.16. If the cooling time for a component exceeds that depicted on the graph, then complete martensite will not be formed and a structural change shown by curves a and b in figure 5.16 will be formed. Because of this slight change in structure from total martensite, a lower hardness value will be developed, hence for maximum hardness the cooling rate is critical.

5.10 Tempering processes

This process is carried out on hardened steels to remove the internal stresses and brittleness created by the severe rate of cooling. Stresses may also be created within a material by matching processes and these may be relieved by tempering. The treatment requires heating the steel to a temperature range of between 200 and 600°C depending upon the final properties desired (figure 5.24). This heat energy allows carbon atoms to diffuse out of the distorted lattice structure associated with martensite, and thus relieve some of the internal stresses. As a result the hardness is reduced and the ductility, which was negligible before treatment, is increased slightly. The combined effect is to 'toughen' the material which is now capable of resisting a certain degree of shock loading. The higher the tempering temperature and greater the capacity for absorbing shock. Figure 5.25 illustrates the temperatures and typical applications of such tempered structures. For a more controlled tempering procedure special tempering furnaces can be used.

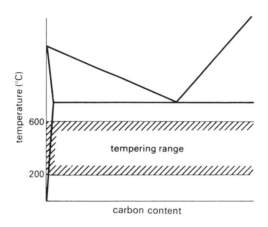

Figure 5.24 Tempering temperature range for iron–carbon alloys.

Tempering temperature (°C)	Oxide film colour	Application
200	—	Scribers
230	Pale straw	Razor blades, woodcutting tools
250	Brown	Punches, reamers
270	Purple	Cold chisels
350	Dark blue	Springs

Figure 5.25 Tempering temperatures and applications

5.11   Nitriding

The process of quench hardening carbon steels involves very rapid cooling rates, which produce the hard martensitic structure. It is possible to produce a hard structure without quenching, one such process is called *nitriding*.

A hard, wear-resistant surface can be produced by using nitrogen instead of carbon. The nitrogen is obtained from ammonia, which 'breaks down' to its constituents of nitrogen and hydrogen when heated to 500°C. The component is heated in an atmosphere of ammonia for between 40 and 90 hours. During the soaking time the ammonia (NH_3) dissociates into $3H + N$ ($NH_3 \leftrightarrows 3H + N$). The atomic nitrogen dissolves into the surface of the iron, which at 500°C is still a BCC arrangement, and is arranged interstitially thus forming a very hard skin. Plain carbon steels are not suitable for nitriding, since the iron nitrides that are produced form a very brittle skin. The most suitable materials contain 1% chromium, 1% aluminium, 0.2% molybdenum and 0.5% carbon. Such alloys are described as *nitralloys*. During the process hard nitrides of aluminium and chromium form in the surface layers. The chromium nitrides diffuse further into the nitralloy, producing a gradual transition from maximum hardness at the skin to a lower hardness value at the core. The effect of molybdenum is to refine and toughen the core.

Since no quenching is required after nitriding, cracking and distortion are unlikely. Surface hardness values of up to 1150 VHN are possible and since the process is conducted at 500°C the hardness values will be maintained up to this temperature. VHN represents a hardness index number obtained by using a square-based diamond pyramid indenter (see section 13.7).

5.12   Mass effects and hardenability

An important factor to be considered when conducting heat-treatment operations, concerns the size of the component. The bigger the component cooled by quenching, the greater the mass of material and the longer it takes for it to attain uniform temperature. This variation in cooling velocity across the section will result in changes in structure and properties. Depth and distribution of hardness throughout the section is defined as hardenability. The mechanical properties obtained by quench hardening a 75-mm diameter bar will be different from those obtained in a 15-mm diameter bar of the same steel composition. Steel manufacturers specify a *ruling section* for particular steel alloys, which is the maximum diameter that will provide the stated mechanical properties after hardening. If the ruling section is exceeded, the stated mechanical properties will not be guaranteed. The mass effects of heat treatment are shown in figure 5.26.

The connection between the temperature, the rate of transformation, the structure obtained and its properties, can be studied by reference to the S curve shown in figure 5.16 or to be more exact a *time temperature transformation curve* or TTT curve.

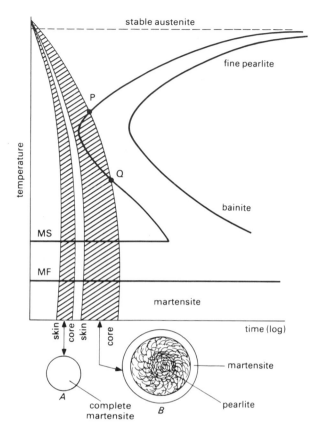

Figure 5.26 Mass effects and hardenability. Specimen A is completely martensitic in structure. Specimen B because of its mass cools more slowly and transformation occurs between P and Q to pearlite. Final structure of specimen B is pearlite core and martensitic skin.

5.13 Time temperature transformation curve construction

These curves show the transition points from one structure to another which occur at certain temperatures on cooling between 723 and 220°C. This temperature range corresponds to the temperatures of the incubation bath, which contains the specimens for varying periods of time, ranging from 1 s up to 1.5 h duration. As shown in figure 5.27, when the 0.8% carbon specimen is held for 1 s in an incubation bath at 500°C, followed by rapid quenching in water, then a martensitic structure is obtained. If incubation time is increased to 6 s, then partial transition occurs and a second structure called bainite is formed alongside the austenite, which, upon quenching becomes martensite with bainite. When the incubation period is further extended to 12 s, complete transition of austenite to bainite occurs, and since no austenite exists, no martensite is formed by the rapid water quench. The resulting structure will be completely bainitic.

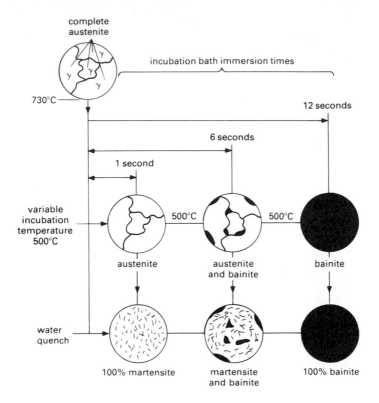

Figure 5.27 The effects of various incubation times at 500°C on 0.8% carbon steel.

The heat treatment sequence connected with TTT curve involves heating the specimen to just above 723°C to obtain stable austenite. The specimen is then quenched in an incubation bath, the temperature of which is variable. After specific times of immersion the specimens are removed and water quenched. The structure obtained is microscopically examined and the degree of structural change can be assessed. When all the results corresponding to the various incubation-bath temperatures and times of incubation are compiled, the TTT curve for that particular alloy can be constructed.

The typical structures that can be obtained for a 0.8% carbon steel alloy are shown in figure 5.28. At about 550°C the critical cooling time is 1 s. To obtain extended incubation times quench the alloy from 723°C to below 550°C in less than 1 s, and then down to 300°C within the remainder of that second. This cooling procedure enables longer soaking times for the component in the incubation bath, *before* transition or structural changes begin. The incubation bath at 300°C allows up to 50 s soaking time for a 0.8% carbon steel component before transition commences.

This extended incubation time enables problems of distortion and quenching cracks to be minimised. Two specific heat-treatment processes that utilise the above quenching methods are *martempering* and *austempering*.

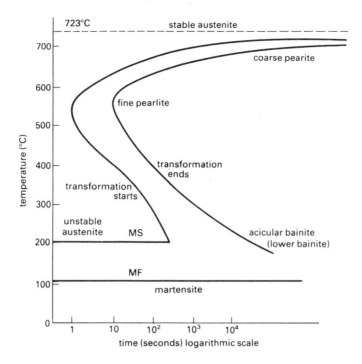

Figure 5.28 Time–temperature–transformation curve for 0.8% carbon steel.

5.14 Martempering

This process involves heating the steel to above 723°C to obtain austenite, it is then quenched rapidly into a molten salt bath maintained at a temperature just above the martensite formation temperature. The steel is held in the incubation salt bath just long enough to enable the structure to attain uniform temperature throughout. It is then withdrawn from the salt bath and allowed to cool slowly to room temperature. The incubation period should not exceed the time where transition or transformation to bainite begins. Because of the limited incubation time this process is limited to relatively small-section work. The cooling period enables both the core and the skin to form martensite simultaneously. The uniform cooling rate reduces both distortion and internal stresses and a uniform martensitic structure is produced (figure 5.29).

5.15 Austempering

The structure obtained in carbon steels by this process is called bainite and is similar to a tempered martensitic structure. Bainite offers the advantage that tempering is not required and thus reduces the number of heat-treatment processes involved. The purpose of austempering is to eliminate the need for

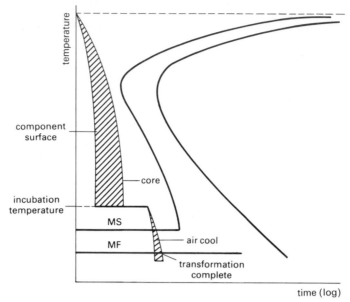

Figure 5.29 Martempering process.

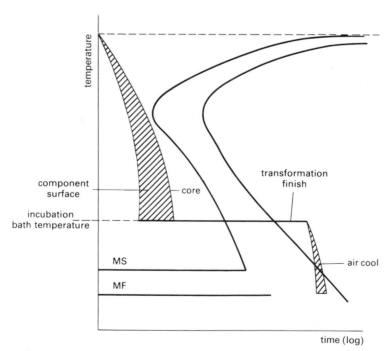

Figure 5.30 Austempering process.

drastic water quenching, thus reducing the effects of thermal shock which could cause cracking and distortion. The toughness and ductility of austempered steel is considered to be better than that of similar steels which have been conventionally quench hardened and tempered.

With austempering it is necessary for the steel component to by-pass the 'nose' section of the curve to ensure transition does not commence prematurely. The component is immersed in an incubation bath which is held at a temperature above the martensite start position. The immersion period is maintained until the component has completely changed to bainite. The component is then cooled to room temperature at any desired rate (figure 5.30).

The principal advantage of the isothermal processes of austempering and martempering is that the possibilities of internal stresses due to rapid quenching are reduced.

Chapter 6

Heat treatment equipment and procedures

The object of heat treatment processes is to alter the structure of the metal either partially or completely. Such procedures or operations ensure that the treated material satisfies the service requirements. All heat-treatment operations involve heating the material by various methods to a given temperature, soaking and then cooling at a pre-determined rate according to the properties required.

6.1   Liquid bath furnaces and applications

These bath furnaces may contain liquid metal or molten salts. The furnace is classed by the reaction of the *liquid* upon the material being treated.

Lead bath

This bath furnace contains molten lead and operates between 400 and 890°C. A neutral atmosphere is produced by this bath furnace, and because lead is a good conductor of heat, uniformity of bath temperature is easily obtained. Lead-bath furnaces are used for simply heating the material to the required temperature without chemical changes occurring in the material. Tempering processes or heating for localised hardening are satisfactorily carried out using the lead-bath furnace. Because of the nature of the fumes given off, good ventilation is required.

Salt bath

Liquid salt baths can be used between 150 and 1350°C depending upon the type of salt selected and the atmosphere required. Salts can provide neutral, oxidising and carburising atmospheres, thus satisfying most heat-treatment requirements. When heated, the salt melts and then acts as a heating medium bringing the immersed component to the required temperature. The salts used are classified as low temperature, medium temperature and high temperature.

A typical salt mix for low-temperature operations such as the tempering process is sodium nitrate 44% and potassium nitrate 56% which melts at 145°C. For medium-temperature operations required for use on non-ferrous metals, potassium chloride 50% and sodium carbonate 50% would be used, melting at 660°C. High-temperature operations include the cyanide hardening process when temperatures around 1000°C are used. For typical applications see figure 6.1. When the component is withdrawn from the salt bath it remains coated

Temperature range (°C)	Salt mixture (%)	Applications
Low temperature 120–650	Sodium nitrate 40–50 Potassium nitrate 50–60	Tempering of carbon steels and low alloy steels
Medium temperature 620–820	Potassium chloride 50 Sodium carbonate 50	Annealing of carbon steels and low alloy steels
High temperature 820–1000	Sodium chloride Barium carbonate 20–30 Sodium cyanide 45–50	Cyanide hardening

Figure 6.1 Salt mixture applications

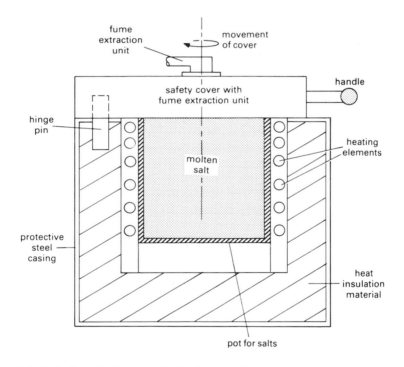

Figure 6.2 Typical salt bath furnace for lead or cyanide.

with a thin protective film of salt until it enters the quenching medium, thus reducing atmospheric contamination. The operator has to be safety conscious and aware of the dangers from fumes, poisoning and explosions created by contact between damp components and the molten salts. A typical salt-bath furnace is shown in figure 6.2.

6.2   Muffle furnace and applications

A furnace, in which the heat source does not directly make contact with the material being treated, is described as a muffle furnace. Articles are heated in a closed refractory lining where the hot gases contact the outside of the heating chamber. The products of combustion do not enter the heating or work chamber therefore its atmosphere can be controlled. This heating arrangement enables the desired reaction at the surface of the component to be controlled. With such a 'muffle' arrangement, scaling of the surface can be prevented. Because muffle furnaces can produce controlled atmospheres, the furnace is suitable for softening of cold-worked metal, or for bright annealing or when it is necessary to prevent decarburisation. Gas-fired and electric muffle furnaces are shown in figures 6.3 and 6.4.

6.3   Non-muffle furnaces and applications

Batch furnace

This furnace is basically a box sealed at one end by a door. The components are charged into the furnace via the door and they remain in position during heat treatment, after which they are removed. This type of furnace, illustrated in figure 6.5, is suitable for small production quantities where annealing, normalising, hardening or pre-heating processes are to be conducted. The heating source consists of coils of nickel–chromium positioned in the hearth roof and side walls.

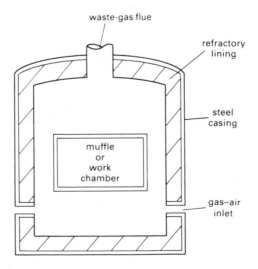

Figure 6.3 Gas-fired muffle furnace.

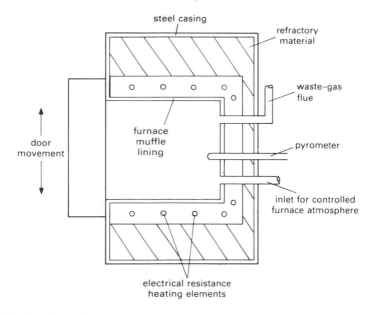

Figure 6.4 Electric muffle furnace.

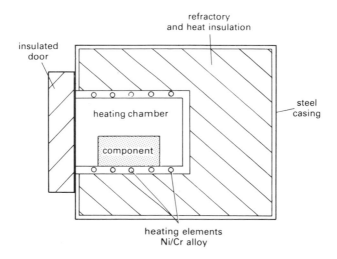

Figure 6.5 Batch furnace.

Natural-draught, gas-fired, semi-muffle furnace

In this furnace combustion occurs beneath the furnace hearth and the products of combustion pass through the work chamber. The heat energy is reflected from the concave roof onto the hearth and the workpiece. Waste gases are removed through a flue positioned in the roof. The combustion chamber and work stage or hearth is surrounded with refractory material as shown in figure 6.6. This furnace is used for normalising, annealing or pack carburising.

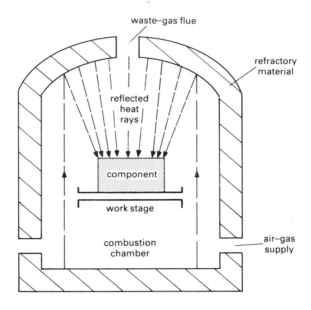

Figure 6.6 Natural draught furnace.

Tempering furnace

The operating temperature of this furnace is relatively low and consists of a chamber in which the air is heated. This hot air is made to circulate by the action of a fan thus ensuring even heating. The furnace is shown in figure 6.7.

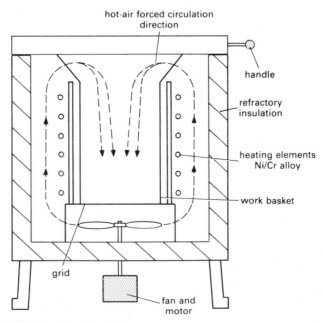

Figure 6.7 Tempering furnace.

6.4 Gas-furnace atmospheres

These may be defined as atmospheres where the composition is deliberately controlled in order to produce a specific effect. The atmosphere produced may be required to prevent or minimise oxidation or produce chemical changes in the surface layers of the steel.

Oxidising reaction

When the temperature of steel is increased, its surface reacts with the surrounding furnace atmosphere. If oxygen is present in excess to that required for combustion in the furnace, oxide scale will form. From such an air–gas mix the products of combustion will be nitrogen (N_2), carbon dioxide (CO_2), water vapour (H_2O) and the excess oxygen (O_2). With the exception of N_2, the other gases, under the influence of an elevated temperature, combine with steel to produce surface oxides. The chemical reaction involving O_2 combined as CO_2, and the ferrite constituent of steel, clearly illustrates the formation of surface oxides,

$$Fe + CO_2 \leftrightharpoons FeO + CO$$

When oxygen in the form of water vapour is present, especially at high temperatures, the oxide-forming reaction with ferrite is

$$Fe + H_2O \leftrightharpoons FeO + H_2$$

Decarburising reaction

The furnace atmosphere may contain elements that have a strong attraction or affinity for the carbon content in the surface of the steel. The reaction of CO_2 apart from forming ferrous oxides, can also combine with the carbon thus decarburising the steel.

$$C + CO_2 \leftrightharpoons 2CO$$

The presence of the oxide scale destroys the surface finish, and the strength and corrosion resistance will also be reduced. When carbon is lost due to scaling the hardening properties are seriously affected.

Reducing–carburising atmosphere

If the oxygen supply in the combustion chamber is reduced, the result is incomplete combustion of the fuel gas. The remaining unburnt gas is described as unburnt hydrocarbon. The products of combustion arising from the gas-rich mixture will be N_2, carbon monoxide (CO), water vapour and the unburnt hydrocarbon. When the hydrocarbon reacts with the surface of the hot steel, oxygen is extracted and carbon is deposited into the surface layers.

Another reducing gas is methane (CH_4). Carbon is released when two molecules of hydrogen are formed (chemically shown as $2H_2$). The liberated carbon can now be absorbed into the surface of the steel. The methane reaction can be shown as

$$CH_4 \leftrightharpoons 2H_2 + C$$

If the carbon monoxide is present in sufficient quantities (chemically shown as 2CO) then this will also lead to carburising. The chemical reaction would be

$$2CO \leftrightharpoons CO_2 + C$$

The released carbon becomes dissolved in the steel. If the furnace temperature is very high some scaling is likely due to the presence of water vapour.

If surface scaling is to be prevented then water vapour must be isolated. This may be achieved by using a muffle furnace (see Section 6.2).

6.5   Quenching media

The purpose of these media is rapid heat removal from the hot immersed component. Because of the infinitely variable component sections, and the differing degrees of structural changes required, various quenching media are used. In order of quenching speeds the most common media are: brine (the fastest), cold water, warm water, mineral oil, animal oil, vegetable oil and air.

Brine

Brine solutions (sodium chloride in water) produce a very drastic quenching reaction. When the hot steel is immersed, a vapour envelope is formed, crystals of salt then form on the steel surface. These salt crystals then explode from the surface taking with them the surface scale. An inrush of cold brine solutions then makes contact with the clean steel and the cooling rate is increased.

Water

Cold-water quenching also produces a vapour envelope which forms an insulating cover between the metal and the water. Agitation of the immersed component will reduce this insulating effect. The result is a slightly less-severe quench. The cooling reactions of brine and water on the outside layers of the immersed steel is very severe, therefore the rate of heat loss is very high. The result is drastic component structural changes, varying in intensity from surface to centre, and in consequence produces severe internal stresses. Such internal distortion can result in warping or formation of quenching cracks. Where distortion must be avoided then brine, or water quenches, should not be used.

Oil

Quenching in oil is less drastic than water thus reducing distortion and warping. To compensate for the slower cooling rate associated with an oil quench, the hardness value is maintained by the presence of alloying elements in the steel.

Before using a quenching oil certain requirements must be satisfied. The oil must be *stable* at the immersion temperature, i.e. it must not oxidise or form a sludge. The flash point of the oil must be high enough to prevent fire risk. The viscosity of the oil must not be too high or oil loss will result from *drag out*, i.e. where the oil adheres to the component.

Air

For certain small intricate sections quenching in oil may be too severe, resulting in warping or cracking. Satisfactory cooling may be achieved by using air. The loss of hardness, associated with such a slow cooling reaction, is overcome by introducing alloying elements to the steel. One such alloy described as an air-hardening alloy would contain 0.45% manganese (Mn), 4.0% nickel (Ni), 1.2% chromium (Cr), and 0.3% molybdenum (Mo).

6.6   Quenching procedure

Many components are rendered useless by excessive warping, or by the formation of quenching cracks. These structural faults arise from unsatisfactory immersion methods or too severe a quenching media. When a suitable media has been selected then consideration to the quenching technique must be given. With cylindrical components vertical quenching is satisfactory (figure 6.8); flat sections should be immersed edgeways; components that are non-uniform in section, should be quenched thick section first. Upon immersion the component should be agitated to ensure even rates of cooling of the component.

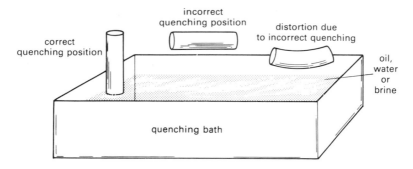

Figure 6.8 Correct quenching procedure.

6.7   Safety

Due to the nature of the equipment, and the operating procedures associated with heat-treatment processes, certain safety attitudes must be developed. The cyanide-salt-bath furnaces must be operated under strict codes of practice. Cyanide salts are extremely poisonous and every precaution must be taken to avoid inhaling the fumes. Fume extraction equipment must operate in unison with the furnace. Protective clothing must be worn by the operator as protection against 'splash back' of the molten salt. This 'explosive' reaction is the result of damp components being immersed in the molten salt. A similar reaction occurs when the components are removed from the salt bath, and quenched in water to produce the hardening reaction. Under no circumstances must the cyanide salt come into contact with an open wound. A protective dressing should be applied. The consumption of food in the cyanide salt bath area of the workshop should also be absolutely forbidden. Medical equipment containing antidotes, must be readily available in the event of an accident involving the cyanide baths. Because of their high operating temperatures, heat treatment furnaces must be operated with care if skin burns are to be avoided. A warning indication must also be made concerning components that have been removed from the furnaces to cool in air. The oxide scale that quickly forms on the hot component can give an incorrect impression of the component temperature and severe burns could result from such a misleading situation. As a result of the Health and Safety at Work Act operatives have certain legal responsibilities concerning their conduct and attitudes to the safety procedures. Apart from their legal responsibilities the operatives must for their own sake be very safety conscious when working with heat-treatment equipment.

Chapter 7

Cast irons

7.1 Definition of cast iron

During the eighteenth and nineteenth centuries cast iron was used extensively for decorative purposes as well as for engineering applications. We have only to look around to see the monuments left to us by the great engineers Telford, Stephenson and Brunel in their development of our transport systems to appreciate their perception in its application and value as an engineering material. It was, and still is the cheapest metallurgical material available. Although its development lagged behind that of steel, considerable metallurgical progress has been made in recent years. It is now extensively used in engineering, second only to wrought steel in terms of annual tonnage produced. Cast iron is now an extremely important engineering material and is suitable for applications hitherto confined to forged steel.

Cast iron is an alloy of iron and carbon, containing between 2.5 and 4.5% carbon. It is basically cupola-refined pig iron to which adjustments in composition have been made during the refining process. Other elements such as silicon, sulphur, manganese and phosphorus may also be present. Cast iron means a wide range of materials of different compositions, structures and hence properties and include irons which are wear, corrosion and heat resistant. Such irons provide ductility and toughness which combines the advantages of cast iron with the strength and toughness of steel.

The various irons can be classified as shown in figure 7.1 based on the form of graphite and the type of matrix structure in which the graphite is contained. It

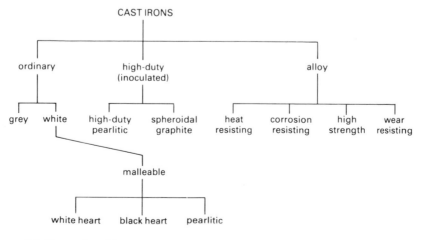

Figure 7.1 Types of cast irons.

91

should be noted that alloy cast irons are not included in this text since they are beyond the scope of the chapter.

The main advantages of cast irons are that they are cheap to produce, strong and rigid in compression, easily machined, have good fluid properties when molten and can be heat-treated to provide toughness and strength.

7.2   Cast iron and the iron-graphite thermal equilibrium diagram

If the iron–iron carbide thermal equilibrium diagram previously considered is extended beyond the *steel* portion, it will be seen that cast iron may be either hypoeutectic or hypereutectic, depending on carbon content. However, when considering cast iron cooled strictly within equilibrium conditions, the high carbon content is present in the form of graphite. Hence consideration of the structure of cast iron must be made against the iron–graphite thermal equilibrium diagram (figure 7.2). This diagram is similar to that for iron–iron carbide except for minor changes in temperature and composition.

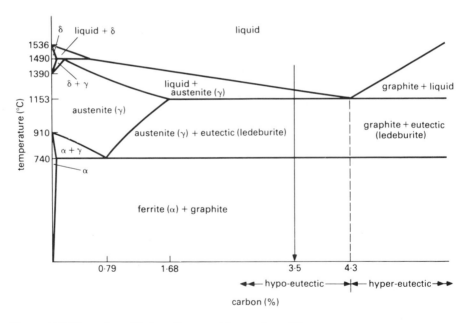

Figure 7.2 Thermal equilibrium diagram of iron and graphite.

From the plain carbon iron equilibrium diagram it can be seen that during cooling from the liquid, through solidification down to room temperature, the cast iron is subjected to a eutectic and eventually a eutectoid reaction. Regardless of carbon content the final structure of the iron will be basically ferrite and graphite, although in practice some carbon forms cementite which combines with ferrite to form pearlite. Such will be the case when the iron–carbon alloy is cooled under equilibrium conditions.

If, however, the rate of cooling is accelerated or other elements are introduced to the iron, then the final structure and hence the properties will differ, particularly in the form and distribution of the carbon.

7.3 Cooling and microstructure of hypoeutectic irons

Consider a hypoeutectic plain carbon iron of, say, 3.5% carbon content, cooling from the liquid state to room temperature under equilibrium conditions (figure 7.2). When the temperature falls below the liquidus, austenite forms dendritically from the liquid, and progresses until the eutectic temperature of 1153°C is reached, with the liquid becoming progressively richer in carbon and its composition following the liquidus line towards the eutectic point. This is the same procedure as described in Chapter 2.

At 1153°C, the remaining liquid, now of 4.3% carbon, is subjected to a eutectic reaction. The result of the reaction in this case is of a different form to that referred to previously. In this high-temperature eutectic reaction the eutectic, known as 'ledeburite', commences to solidify from nuclei from each of which a spherical solid is formed. In these 'eutectic cells' growth of austenite and graphite is simultaneous. Both constituents are in continuous contact with the liquid and 'feed' from it. The graphite, in fact, forms a branched skeleton similar in appearance to a rosette, within the austenite (figure 7.3). It should be noted that the graphite rosettes develop three-dimensionally.

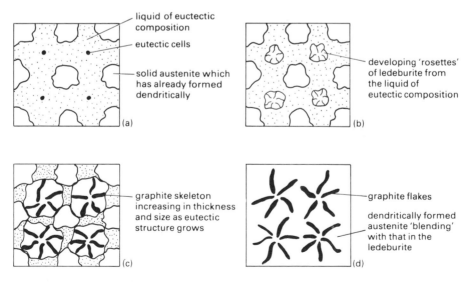

Figure 7.3 Diagrammatic representation of the formation of the eutectic ledeburite in a hypoeutectic cast iron. (a) Formation of eutectic cells. (b) Development of ledeburite (c) Eutectic reaction nearing completion. (d) Solidification complete, with graphite rosettes in a matrix of solid-solution austenite.

On completion of the eutectic reaction and, in fact, solidification, the austenite has a composition of 1.68% carbon. With subsequent cooling the solubility of the austenite decreases, following the A_{cm} line towards the eutectic point. This causes further graphite to be rejected which combines with, and increases the size of the rosettes. At 740°C the austenite, of 0.79% carbon content, would theoretically transform to ferrite and further graphite during the eutectoid reaction. However, due to the sluggishness of the reaction, pearlite—laminations of ferrite and cementite—is formed. The resulting structure would be one of graphite 'rosettes' in a pearlitic matrix (figure 7.4).

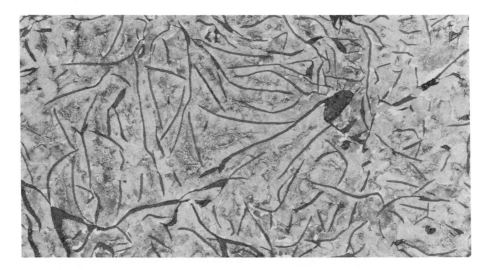

Figure 7.4 Final structure of a slowly cooled hypoeutectic iron (grade 150 ordinary cast iron). (Courtesy of BCIRA).

7.4 Effect of cooling rate and section variation on structure

In order that the structure at any given temperature can be predicted, reference can be made to the iron–graphite thermal equilibrium diagram (figure 7.2). In such cases the cooling rate must be slow enough to maintain equilibrium conditions, giving sufficient time for the carbon to form as graphite and so produce grey iron. However, if the cooling rate is accelerated, less time will be available for the formation of the graphite and iron carbide will form producing a white iron. In this case the resulting structure will be one of pearlite in a network of iron carbide (figure 7.5).

As the cooling rate is retarded the iron carbide becomes unstable and some decomposes to form graphite. The slower the cooling rate, the more unstable the iron carbide becomes and hence the more graphite is produced. This will produce a structure of small graphite flakes in a matrix of fine pearlite (figure 7.6). Such an iron is often referred to as mottled iron.

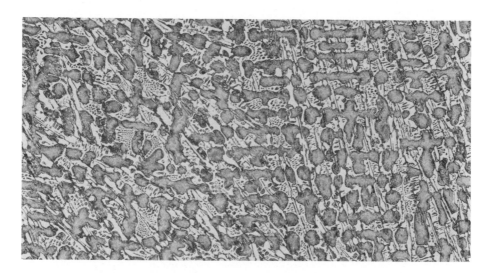

Figure 7.5 Typical white cast iron. (Courtesy of BCIRA).

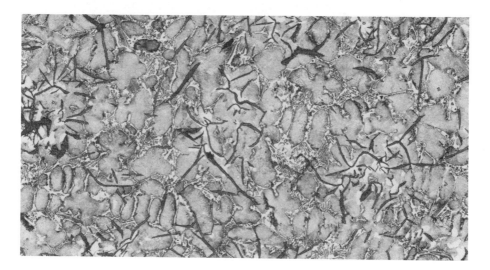

Figure 7.6 Typical mottled cast iron. (Courtesy of BCIRA).

The descriptions of all three irons, as grey, white or mottled is derived from the appearance of the fracture. A structure containing free graphite will give a grey appearance, one containing primary cementite a white appearance and a structure containing a mixture will appear somewhere between the two others.

If an iron casting is cooled at industrial cooling rates, the resulting structure across the section of the casting will vary, from white iron at the outside to grey iron at the core. This feature is readily demonstrated in a casting of conical shape, a diagrammatic representation of the structural variation across the section being illustrated in figure 7.7.

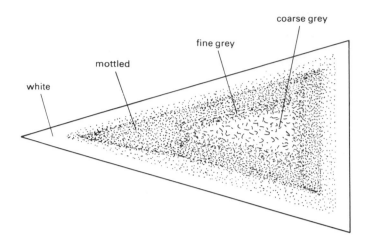

Figure 7.7 Section across a conical-shaped ordinary iron casting.

The outer areas of white iron will be very hard and difficult to machine. This is a result of the rapid cooling due to the chilling action of the mould which prevents the decomposition of the cementite. However, towards the centre the iron cools more slowly, allowing a progressive increase in decomposition. This results in more graphite forming and a softer and more easily machined iron. At the core, the coarse grey iron will exhibit large flakes of graphite in a ferrite matrix.

Although the cooling rate and hence the structure can be controlled within limits, the shape of the casting will determine the final structural variation. This can be improved by controlling the composition, particularly the elements such as silicon and sulphur.

7.5   Influence of composition on structure

Cast iron is basically an alloy of iron containing up to a total of 10% of the elements of carbon, silicon, sulphur, manganese and phosphorus.

Carbon

In cast iron carbon can exist in two forms: free graphite, or combined with some iron to form iron carbide or cementite. The form in which carbon exists depends on the presence of silicon and sulphur and the rate of cooling. Cementite, being hard and brittle, is predominant in white iron whereas grey iron contains graphite. In addition, carbon influences the melting temperature of the iron. As the carbon content increases the temperature reduces to a minimum at the eutectic composition (figure 7.2).

Silicon

An element which influences the form which carbon takes, silicon renders the cementite unstable so that it decomposes, forming ferrite and graphite, thus producing grey iron. As the silicon content is increased, the stress-raising graphite flakes become coarser. However, if the silcon content is in excess of that required for the decomposition of all the cementite, the remainder will dissolve in the ferrite and cause an increase in strength, hardness, and brittleness. On the other hand, since more ferrite is produced by the excess silicon a certain amount of softening is induced together with an increase in toughness. Another major and beneficial effect of silicon is that it increases the fluidity of the molten iron thus improving the casting properties.

Sulphur

This has the opposite effect to that of silicon in that it tends to stabilise the cementite. It inhibits graphitisation and produces a white iron which is hard, brittle and difficult to machine. In addition, its presence in the form of iron sulphide (FeS) tends to promote brittleness.

Manganese

The presence of undesirable sulphur is controlled by the use of manganese which combines with sulphur to form insoluble manganese sulphide (MnS), which floats to the top of the molten iron and combines with the slag. This indirectly promotes graphitisation as the sulphur content is reduced. However, manganese also has a stabilising effect on cementite in its own right which offsets the graphitising effect resulting from the reduction in sulphur content. The more direct effects of manganese are to harden the ferrite constituent in which it is soluble, to stabilise cementite and so pearlite and to promote a fine grain structure. All of these effects lead to an increase in strength. The manganese content in cast iron is calculated from the formula

$$\% \, Mn = (1.7 \times \% \, S) + 0.3.$$

Phosphorus

This is present in cast irons as iron phosphide (Fe_3P), which forms a eutectic with ferrite in grey irons and with ferrite and cementite in white irons. Since these eutectics have a considerably lower melting temperature than the normal eutectic—950°C as compared with 1153°C—high-phosphorus irons have greatly improved fluidity. Such irons are ideally suited to castings having thin sections. Unfortunately only a small amount of phosphorus produces a large volume of phosphide eutectic which is hard and brittle. For this reason its content must be kept down to a low level when the property requirement of the casting is toughness.

   Both *silicon* and *phosphorus* have an almost identical influence on the carbon content of the eutectic ledeburite. Additions of silicon to the iron lower the percentage carbon content in the eutectic by 0.33% for each 1% silicon present, with phosphorus having a similar effect. Hence, for an iron required for intricate castings and having 2% silicon and 1% phosphorus content, the eutectic would have an equivalent carbon content of 3.3%. This is known as the *carbon equivalent value* (CEV).

$$CEV = total\ \%C + \left(\frac{\%Si + \%P}{3}\right).$$

In order that a cast iron containing silicon and phosphorus can be considered as a simple binary alloy, the CEV is plotted against temperature as in any thermal equilibrium diagram.

7.6   Ordinary cast irons

There are two types of ordinary cast iron—grey and white—which have approximately the same carbon content of 3.25%. Their difference lies in the silicon and phosphorus content, both of which influence the form which the carbon takes. It has previously been stated that silicon is a strong graphitiser, promoting the breakdown of cementite into its constituents of ferrite and free graphite in the form of flakes. Hence, irons with a low silicon content will produce a white iron and those with a high silicon content produce a grey iron. In addition, high cooling rates prevent the decomposition of cementite thus producing a white iron. If moderate cooling rates are applied some combined in the form of cementite will not be broken down and an iron having both combined carbon as cementite (in pearlite) and free carbon (as graphite) will result producing a mottled iron.

For a cast iron of 3.25% carbon content produced under normal industrial cooling rates it will be the silicon content which influences the structure. Such irons could range from a white iron in which all the carbon is combined to a ferritic iron in which all the carbon is free. This is illustrated by relating the silicon content in the iron to the eutectic composition of 4.3% carbon (figure 7.8).

Grey cast irons

Grey cast irons to BS 1452 are essentially alloys of iron, carbon and silicon. Their structure ranges from graphite flakes in a pearlitic steel-like matrix to one of graphite flakes and areas of pearlite in a ferritic matrix. Of the seven grades shown in BS 1452, grades 300, 350 and 400 have additional alloying elements present which provide high strength at the expense of machining properties. Such grades are often referred to as *high-duty grey irons* although strictly they are *alloy cast irons*. Typical mechanical properties are shown in figure 7.9.

Since the structure of grey cast iron can vary together with the silicon and phosphorus content, the mechanical properties will vary. Such variations will influence application. Where maximum strength is required the carbon content will be lower, reducing the amount of free graphite in the pearlitic matrix. The phosphorus content will also be low, minimising the presence of the phosphide eutectic. If good machining properties are required it will be at the expense of strength. This would be provided with a higher carbon content producing more free graphite in a ferritic matrix. However, where improved casting properties are

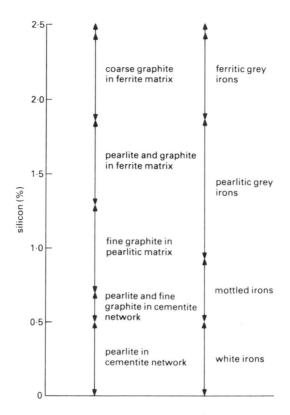

Figure 7.8 Normal industrial cooling of 3.25% carbon ordinary cast irons.

required, the carbon equivalent value should be near to the eutectic composition to reduce the melting temperature. Reduction in the melting temperature is also influenced by an increase in the phosphorus content.

Applications cover a wide range, utilising the low casting temperature (1130–1250°C), low solidification shrinkage (0.5–2.0% as compared with 5% for steel) and good fluidity and damping properties, which illustrate the versatility of the material. Such irons are used for ingot moulds and furnace components, engine cylinder blocks and exhaust manifolds, brake drums and friction pads and machine-tool beds.

White cast iron

White iron, with its structure of primary cementite in a pearlitic matrix has limited application. Its hardness and associated machining difficulties make it truely an 'as cast' material. Applications are in components where severe abrasive conditions are found and in crushing and pulverising equipment.

However, the instability of its structure when subjected to high temperatures is taken advantages of in producing a 'modified' high-duty cast iron. Such irons are produced by a *malleablising* process.

Grade	Tensile strength (N/mm^2)	Compressive strength (N/mm^2)	Elongation (%)	Impact strength (J)	Structure	Comments
150	150	600	0.60–0.75	8–13	Coarse graphite flakes and pearlite in ferritic matrix	Available with high phosphorus content
180	180	672	0.50–0.70	8–13	Medium graphite flakes in pearlitic matrix	Available with high phosphorus content
220	220	768	0.39–0.63	8–16	Medium graphite flakes and ferrite in pearlitic matrix	Available with high phosphorus content
260	260	864	0.57	13–23	Medium graphite flakes in pearlitic matrix	Available with high phosphorus content
300	300	960	0.50	16–31	Fine graphite flakes in pearlitic matrix	Additional alloying elements present
350	350	1080	0.50	24–47	Fine graphite flakes in pearlitic matrix	Additional alloying elements present
400	400	1200	0.50	24–47	Fine graphite flakes in pearlitic matrix	Additional alloying elements present

Figure 7.9 Typical mechanical properties of ordinary grey irons

7.7   Malleable cast iron

In order to provide a cast iron which has similar properties to steel, that is, both strong and ductile with good impact properties, the structure must take a more refined form. The graphite flakes must be smaller and more evenly dispersed throughout the structure. Such refined structures are possible by *heat treating* white iron.

Malleable cast iron was first produced in to Europe in the early eighteenth century. Thin sections of the decarburised malleable iron produced displayed a white appearance when fractured due to a completely ferritic structure and so was described as *whiteheart*. The original *blackheart* or ferritic malleable irons were first produced in the USA in the 1830s in attempts to reproduce the European or whiteheart process. Such irons were so described from the influence of the free graphite in the ferritic matrix in the fracture. These early malleable irons are considered as the forerunners of steel since they can be cast into shape and forged as necessary after suitable heat treatment.

In addition to the whiteheart and blackheart malleable irons there is also available today a pearlitic malleable iron. This iron is harder and stronger and can compete on equal terms with steel. Over the last three decades the malleable irons collectively have emerged as a major engineering material.

The composition of white iron used for malleablising processes has a low silicon and phosphorus content. Typical compositions are shown in figure 7.10. The white irons, when subjected to the appropriate heat treatment produce graphite as temper carbon *rosettes*, roughly spherical in shape, with none of the stress-raising characteristics of the flakes found in grey iron. This refinement produces improved irons with up to 700 N/mm^2 tensile strength and ductility as high as 12% elongation.

Type	Composition (%)				
	C	Si	Mn	S	P
Whiteheart	3.5	0.65	0.5	0.25	0.1 max
Blackheart Pearlitic	2.5	1.0	0.4	0.08	0.1 max

Figure 7.10 Typical compositions of white iron for malleabilising

Whiteheart malleable cast iron

To produce whiteheart malleable iron, white iron of the composition shown in figure 7.10 is packed into steel containers with an oxidising material, hematite ore (Fe$_2$O$_3$), sealed and heated at 950–1000°C for 25–70 h (i.e. austenised). At this temperature the pearlite in the white iron transforms to austenite and eventually the primary cementite dissolves into the austenite. The austenite becomes carbon saturated which results in the rejection of temper carbon in the form of spheroids. Simultaneously, surface decarburisation takes place which results from carbon at the surface of the iron oxidising when in contact with the hematite ore and being lost as carbon dioxide. This causes more carbon to diffuse outwards from the core which in turn is lost by oxidation. The loss in carbon is compensated for by the higher carbon content of the white iron. The structure is now one of austenite of 1.3–1.5% carbon with temper carbon in the core. As a result of surface decarburisation to zero carbon content there is a carbon gradient from the core to the outer surface. The iron is then slowly cooled to a subcritical temperature of 500 to 600°C within the furnace, followed by air cooling. During the slow furnace cooling the austenite with the higher carbon content transforms to pearlite leaving the surface areas as ferrite. Hence, as a result of graphitisation at the core and decarburisation at the surface, the final structure is one of relatively soft, ductile surface layers of ferrite surrounding a tough, high tensile core of temper carbon rosettes in a pearlitic matrix. Figure 7.11 illustrates a typical structure.

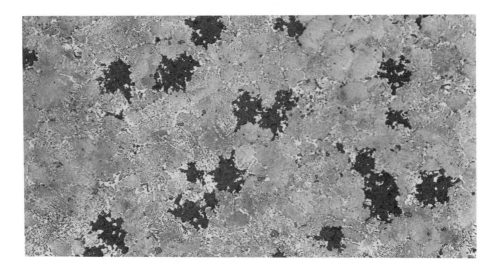

Figure 7.11 Structure of the core of a whiteheart malleable iron casting. (Courtesy of BCIRA).

British Standard 309 lists two grade of whiteheart malleable iron, identified as W410/4 and W340/3. The letter identifies the type of iron, the first numbers identify the minimum tensile strength and the last number refers to the minimum percentage elongation. Typical properties of whiteheart malleable irons are shown in figure 7.14. Applications include pipe fittings and couplings, agricultural and textile machinery parts, electrical switchgear parts, car components and steering boxes for commercial vehicles and components in domestic appliances.

Blackheart malleable cast iron

The production of blackheart malleable iron is similar to that for whiteheart iron except that decarbursing is minimised. In consequence the carbon content of the white iron is lower (figure 7.10). Originally the white iron was packed and sealed in containers, but in this case with a non-reactive material, and austenised at temperatures between 850 and 950°C for up to 100 h. However, development of modern furnaces obviates the necessity to pack in an insulating material since heating and cooling times have been reduced. Now, malleablising can be achieved in less than 50 h. Such furnaces are of the continuous type in which a controlled non-oxidising atmosphere is circulated.

During the annealing process the cementite again goes into solution in the austenite but in this case, when the austenite becomes saturated the carbon is subsequently precipitated in the form of small rosettes of temper carbon. The resulting austenite of 1.1–1.3% carbon precipitates further temper carbon during the very slow cooling below the eutectoid temperature. The final structure is one of evenly distributed temper carbon rosettes in a ferrite matrix, a material which is soft, easily machined and almost a ductile as steel (figure 7.12).

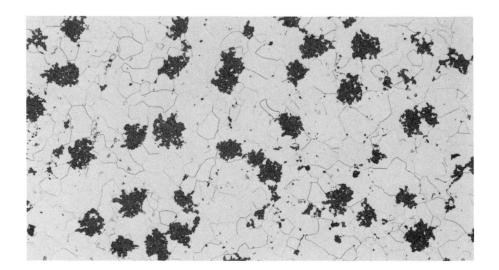

Figure 7.12 Structure of blackheart malleable cast iron. (Courtesy of BCIRA).

British Standard 910 lists three grades of blackheart malleable iron, identified as B340/12, B310/10 and B290/6, the identification having the same interpretation as for whiteheart malleable irons. Typical properties of these irons are shown in figure 7.14. Industrial applications are automobiles, particularly in suspension and transmission components, and agricultural and earth-moving machinery.

Pearlitic malleable cast irons

This malleable iron is produced from white iron of similar composition to that used in the whiteheart process, with similar heat treatment. However, the heat treatment is controlled by initially malleablising, partially or completely at 950°C to bring about adequate breakdown of the cementite. It is then reheated to 900°C, quenched and, if necessary, tempered to produce a matrix of pearlite or tempered

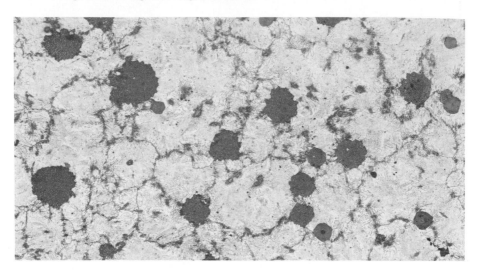

Figure 7.13 Structure of pearlite malleable cast iron. (Courtesy of BCIRA).

martensite instead of ferrite. Alternatively pearlitic malleable iron can be produced by increasing the carbide-stabilising element, manganese which causes a retention of the pearlitic matrix (figure 7.13).

British Standard 3333 lists five grades, P690/2, P570/3, P540/5, P510/4, and P440/7. Again the identification has the same interpretation as for the other malleable irons, typical properties being illustrated in figure 7.14. No impact value is quoted since this can vary with tempering.

Type	Grade	Tensile strength (N/mm^2)	Elongation (%)	Impact strength (J)	Hardness (H$_B$)
Whiteheart	W410/4	410	4	50	120–229
(BS 309)	W340/3	340	3	50	120–229
Blackheart	B340/12	340	12	14–18	110–149
(BS 310)	B310/10	310	10	14–18	110–149
	B290/6	290	6	14–18	110–149
Pearlitic	P690/2	690	2	—	241–285
(BS 3333)	P570/3	570	3	—	197–241
	P540/5	540	5	—	179–229
	P510/4	510	4	—	170–229
	P440/7	440	7	—	149–197

Figure 7.14 Typical properties of malleable cast irons

7.8   Inoculated cast irons

Although the shape and distribution of the free graphite can be altered by the lengthy malleablising process an alternative method of achieving similar results is by *inoculation* i.e. by adding other substances to an existing molten alloy. Molten irons are inoculated with a graphitising agent just before casting, which achieves greater graphitisation. Such irons, with a finer and more evenly dispersed graphite content possess improved properties. Basically there are two types of inoculated iron, the pearlitic high-duty irons of the meehanite and ni-tensyl types and spheroidal graphite cast iron.

Pearlitic high-duty cast iron

This iron is produced by inoculating a superheated ordinary white iron just prior to casting. The inoculant produces a large number of small nuclei which, in turn, promote the formation of fine, evenly distributed graphite flakes. One such iron, marketed under the trade name *meehanite* is inoculated with calcium silicide. Another iron, *ni-tensyl*, is inoculated with nickel shot and ferro-silicon. In both cases fine graphite flakes are produced, evenly distributed in a pearlitic matrix. Typical compositions and properties are shown in figure 7.15.

Type	Composition (%)						UTS	Hardness
	C	Si	Mn	S	P	Ni	(N/mm^2)	(H_B)
Flake graphite 'meehanite'	2.5	1.3	0.8	0.14	0.1	—	150–480	150–280
Ni-tensyl	2.8	1.5	0.8	0.6	0.1	1.5	340–460	280

Figure 7.15 Typical composition and properties of inoculated ions

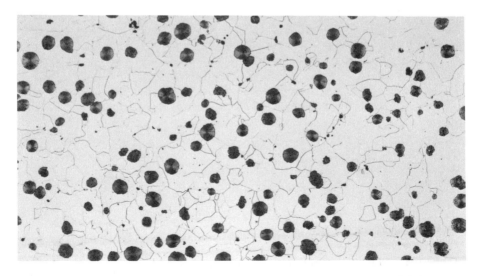

Figure 7.16 Spherical graphite iron with a fully ferritic matrix. (Courtesy of BCIRA).

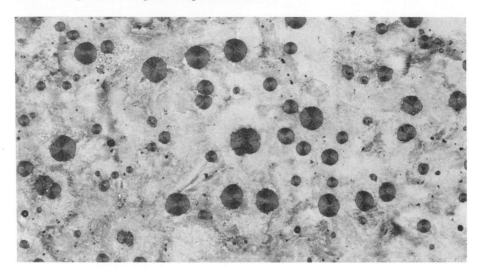

Figure 7.17 Spherical graphite iron in an 'as cast' structure. (Courtesy of BCIRA).

Spheroidal graphite cast iron

Spheroidal graphite (SG) iron is sometimes known as *nodular* or *ductile* cast iron and is now the most important cast iron used. It is produced by inoculating the iron with a carbide-stabilising element such as magnesium or cerium just prior to casting. Magnesium, in the form of a nickel-magnesium alloy, is more widely used. The resulting iron can be heat treated to vary the structure and so the mechanical properties. A relatively high-carbon and silicon content (CEV approximately 4.3%) ensures good castability in all shapes and sizes. Since SG iron has a melting temperature and fluid characteristics similar to grey iron, there are obvious casting advantages compared with cast steel. The resulting structure shows a spheroidal graphite formation which has no stress-raising characteristics in a matrix which will vary with heat treatment. Figure 7.16 and 7.17 illustrate the structures of SG irons with a fully ferritic and mainly pearlitic matrix.

British Standard 2789 covering SG irons lists six grades ranging from irons with a ferritic matrix to irons with a tempered martensitic matrix. Figure 7.18 lists typical properties together with structure and appropriate treatment.

Grade	Tensile strength (N/mm²)	Minimum elongation (%)	Impact strength (J)	Hardness (H$_B$)	Normal matrix	Normal heat treatment
370/17	370	17	12 (minimum)	180	Fully	900°C for 2–4 h
420/12	420	12	8	200	Mainly	700°C for 6–12 h
500/7	500	7	4	170–240	Ferritic/pearlitic	900°C for 2–4 h furnace cool
600/3	600	3	1.4	190–270	Mainly pearlitic	'As cast' condition
700/2	700	2	2.7	230–300	Fully pearlitic	850–900°C for 2–4 h air cool
800/2	800	2	2.7	250–350	Tempered martensite	850–900°C for 2–4 h quench and temper

Figure 7.18 Typical properties and treatment of spheroidal graphite irons

Chapter 8

Light alloys

Light alloys are becoming increasingly important to the engineer as aerospace and energy conservation development proceeds. If alloys of low relative density but possessing high strength are available, a smaller amount of energy would be required to propel a vehicle constructed of such an alloy. This is currently the case with aerospace and land vehicles.

Alloys of both aluminium and magnesium have been developed with high strength: weight ratios, their high strengths being achieved through cold working or heat treatment.

8.1   Heat treatment of light alloys

Light alloys capable of responding to heat treatment must contain elements which have complete solubility in the liquid state and limited solubility in the solid state. The solid solubility should increase with an increase in temperature. Figure 8.1 illustrates the appropriate portion of a typical alloy system on which is superimposed a suitable composition.

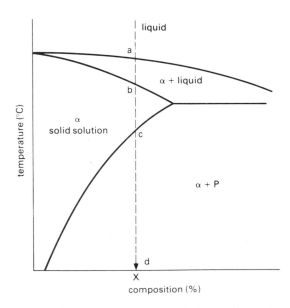

Figure 8.1 Alloy systems which possesses limited solid solubility, the solubility increasing with an increase in temperature.

As an alloy of composition X cools under equilibrium conditions from a to b the α-phase solidifies dendritically in the cooling liquid. When the temperature has fallen to b, solidification will be complete and the alloy will consist entirely of α-phase solid solution. Between temperatures b and c the solubility of the α-phase initially increases following the solidus to the eutectic temperature, and then decreases following the solvus to room temperature. At temperature c the alloy is at saturation point and cannot contain any more alloying element. Since there will be a continued fall in temperature beyond this point there will be a progressive precipitation of the alloying element. However, normal industrial cooling rates are higher than that required for equilibrium. This will give insufficient time for the excess alloying element to precipitate and diffuse to the grain boundaries and hence will be trapped within the lattice structure of the crystals or grains of the α-phase. In some alloy systems the precipitant may be an intermetallic compound.

Alternatively, if the alloy is quenched from a temperature within the α-phase down to room temperature a supersaturated solid solution will result. Precipitation by diffusion will then take place slowly and naturally at room temperature over a period of time.

This is the basis of *precipitation hardening* of light alloys.

8.2 Pure aluminium

Aluminium was first produced in 1825 by the Danish scientist H. C. Oersted but his method of production was very expensive. It was not until 1886 that aluminium was extracted from the ore bauxite by the cheaper electrolytic process. However, the metal did not find serious engineering application until the First World War, following the German discovery of age-hardening an alloy of aluminium and copper. The first significant application was in the structural members of the Zeppelin airships.

Aluminium has a high affinity for oxygen and for this reason it cannot be fire-refined. However, in practice, it has a high resistance to atmospheric oxidation (corrosion). This is the result of the very thin, dense oxide film which forms on the surface and prevents further oxygen from penetrating through the film, sufficient to protect the metal from further atmospheric attack. A process known as 'anodising' has been developed which artificially thickens the oxide skin.

Its high thermal conductivity together with good corrosion resistance make aluminium and its alloys highly suitable for the manufacture of cooking utensils. It also finds application in the production of thin foils and disposable, collapsible tubes for foodstuffs and toilet preparations. Such applications utilise the high malleable properties (resulting from its FCC lattice structure)

Although the electrical conductivity is only half that of copper, taken weight-for-weight for the same length of material, it is a better conductor. This is due to the fact that its relative density is 2.7—approximately one-third that of copper. The relative density also provides a very good weight : strength ratio when compared with steel. Since iron has a relative density of 7.9 and its ultimate strength is about 350 N/mm^2 as compared with 60 N/mm^2 for pure aluminium, an equal weight of aluminium will possess approximately one-half the strength of iron. However, as alloying elements are added to iron to improve strength, they are also added to aluminium, such additions providing some aluminium alloys with an ultimate strength as high as 530 N/mm^2 after heat treatment.

8.3   Elements alloyed with aluminium and their effect

It can be seen that although pure aluminium has certain physical advantages, the mechanical properties are inadequate for most engineering purposes. These inadequacies can be overcome by alloying certain elements with pure aluminium, the most common alloying elements being magnesium, copper, manganese, silicon and zinc.

Magnesium

Resistance to corrosion by seawater and marine atmosphere of cold-worked alloys is improved by the addition of magnesium. However, high percentages reduce the hot- and cold-working properties and so a compromise must be made in deciding the amount to be added.

Copper

Addition of copper to improve the hardness and strength of aluminium alloys is achieved through heat treatment and the formation of the intermetallic compound $CuAl_2$.

Manganese

By promoting the formation of a fine grain structure manganese is used to improve the tensile strength. This improvement is achieved without seriously affecting the working properties.

Silicon

This is an important element in alloys used for castings. Its effect is to improve the fluidity of the alloy and so enable the casting of more intricate shapes. Very high content—over 16%—imparts a very low coefficient of thermal expansion with a high hardness value.

Zinc

Significant in aerospace alloys, zinc improves the strength of aluminium nine-fold. Alloys containing up to 6% zinc can develop an ultimate strength of 530 N/mm^2 after heat treatment so giving a high strength weight ratio.

As an alternative to, or in support of copper, silicon and magnesium can be used in any combination to improve hardness and strength. In this case the magnesium and silicon combine to form the intermetallic compound Mg_2Si.

Other elements used in aluminium alloys include nickel, titanium, iron, chromium and tin.

8.4   Heat treatment of aluminium alloys

Alloys such as those of aluminium and copper respond to heat treatment as described in Section 8.1, since they possess limited solid solubility and the solubility increases with an increase in temperature. Figure 8.2 illustrates the aluminium-rich part of the aluminium–copper thermal equilibrium diagram, which also indicates the precipitate as the intermetallic compound $CuAl_2$.

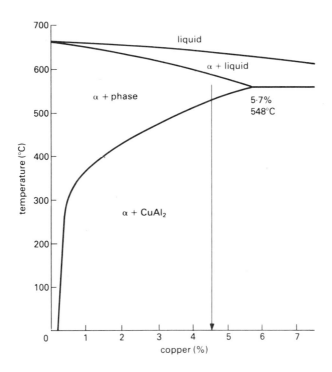

Figure 8.2 Aluminium-rich part of the copper–aluminium equilibrium diagram.

For an aluminium–copper alloy to respond to heat treatment the amount of copper contained in the alloy must be less than the maximum amount which can be contained by the solid solution, i.e. 5.7% at 548°C.

Consider an aluminium alloy containing 4.5% copper which has been raised to a temperature within the α-phase, i.e. 550°C. If the alloy is held at this temperature for a sufficient duration, all the copper will be taken into solid solution—a structure which is soft and ductile. If, however, the alloy is now quenched to room temperature there will be insufficient time for any precipitation and a super-saturated solid solution will result. This structure is soft and ductile, produced by the process known as solution treatment.

The supersaturated condition is unstable and the excess alloying element above 0.2% will eventually be precipitated out of the solid solution as the alloy achieves equilibrium conditions. In some cases precipitation occurs naturally at room temperature when it will be found that hardness and strength increases slowly, reaching a maximum in up to 7 days. Due to the low-temperature precipitation it is sub-microscopic but the changes in the mechanical properties indicate that the intermetallic compound is trapped within the aluminium lattice, increasing the strength and hardness by making slip more difficult. This is known as *natural ageing* or *age-hardening*.

Alternatively, if the alloy does not respond to natural ageing, precipitation of the excess alloying element can be brought about or accelerated by raising the temperature to approximately 160°C and soaking for some 10 h, followed by air cooling. Suitable alloys subjected to this treatment achieve maximum properties (figure 8.3). This treatment is known as *precipitation hardening*.

From figure 8.3 it can be seen that if the alloy is held at the precipitation-hardening temperature for too long, the properties will fall dramatically and the precipitate will be enlarged and microscopically visible. When this situation arises, the alloy is said to be *over-aged*. Hence, time and temperature for precipitation hardening are critical.

Aluminium alloys containing appropriate amounts of silicon and magnesium can be treated in a similar manner. In this case the precipitate is an intermetallic compound of the two elements, forming Mg_2Si.

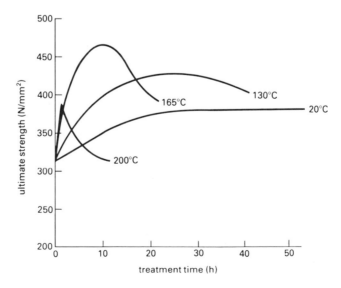

Figure 8.3 Relationship between time and temperature of precipitation treatment on the ultimate strength of a typical aluminium alloy.

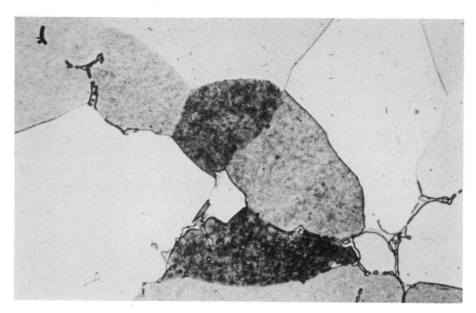

Figure 8.4 Micro-structure of a solution-treated aluminium–4% copper alloy. The structure consists of equi-axed grains of the solid solution of copper in aluminium with no evidence of any precipitation.

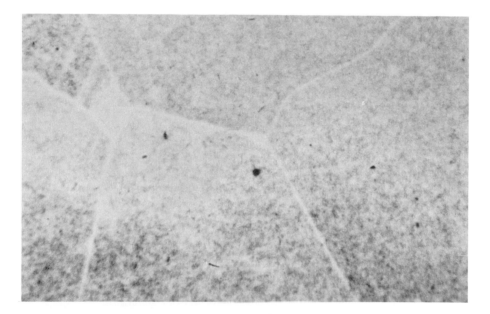

Figure 8.5 Micro-structure of a severely over-aged aliminium–4% copper alloy. In the structure the light coloured precipitate is now microscopically visible with a greater degree of precipitation at the grain boundaries.

It follows that if raising the temperature of solution-treated aluminium–copper alloys will accelerate precipitation then refrigeration will retard it. Precipitation can be delayed for about 4 days by refrigeration at temperatures of between -5 and $-10°C$, an application being for rivets in the aerospace industry.

8.5   Aluminium alloys

Aluminium alloys can be conveniently divided into two groups, wrought alloys and cast alloys. Both of these groups can be further divided into alloys which respond to heat treatment and those which do not.

In order to identify these alloys, the British Standards Institute have produced specifications which indicate the nomenclature to be used. The current nomenclature (revised 1980) is briefly as follows.

Casting alloys
(to BS 1490)

All alloys carry an LM designation which is a prefix to the number which identifies it in the series. The BS identification also carries a suffix which is the temper designation and describes the condition of the casting. The temper designations are: M, as cast with no further treatment; TS, stress-relieved only; TE, precipitation treated; TB, solution treated; TB7, solution treated and stabilised; TF, solution treated and precipitation treated; TF7, solution treated, precipitation treated and stabilised.

The casting alloy series is conveniently subdivided into two groups, *general-purpose alloys* and *special-purpose alloys* which aids the selection of an alloy for a particular application.

Example: LM 10–TB is a solution-treated 10% magnesium special-purpose alloy with high strength and shock resistance and with high corrosion resistance.

Wrought alloys
(to BS 1470, for plate sheet and strip; BS 1471, for drawn tube; BS 1472, for forging stock and forgings; BS 1473, for rivet, bolt and screw stock; BS 1474, for bars, extruded round tube and sections; BS 1475, for wire)

Of the 18 wrought alloys included in the above six standards not all are available under every specification. Wrought alloys now carry the international alloy designation of four digits, the first of which indicates the group of alloys by composition, i.e. aluminium (99.00% minimum or greater) 1* * *, copper 2* * *, manganese 3* * *, silicon 4* * *, magnesium 5* * *, magnesium and silicon 6* * *, zinc * * *, other elements 8* * *, unused series 9* * *. The second digit in the designation indicates alloy modification, zero indicating the original alloy. The last two of the four digits serves to identify the alloy within the group. Within the individual BS specifications the alloys are sub-divided into unalloyed, non-heat-treatable and heat-treatable groups. The alloys also carry a temper designation as

Type	Designation	Basic composition					Treatment	Properties			Applications
		Cu (%)	Mn (%)	Si (%)	Mg (%)	Zn (%)		0.2% proof (N/mm²)	UTS (N/mm²)	Elongation (%)	
BS1470 to BS1475	6063			0.4	0.7		TF	160	185	8	Extruded section
	2031	2.3		0.9	0.9		TF	300	385	6	Forging
	2041A	4.5	0.8	0.7	0.5		TB	—	385	—	Rivet stock
	2618A	2.2		0.2	1.5		TF	340	430	5	Forging
	6061	0.3		0.6	1.0		TF	225	295	9	Drawn tube
	6082		0.7	1.0	0.9		TB	120	200	15	Sheet
Aerospace 'L' series	3L63	4.5	0.8	0.7	0.5		TF	370	450	7	Drawn tube
	2L84	1.5		1.0	0.8		TB	190	310	13	Extruded section
	2L93	4.5	0.8	0.7	0.5		TH	410	460	7	Plate
	2L95	1.6			2.5	5.8	TH	450	530	8	Plate
	2L97	4.3	0.6		1.5		TD	280	430	10	Plate
	L102	4.4	0.8	0.7	0.5		TB	270	410	8	Extruded section
	L113		0.7	1.0	0.8		TF	255	295	8	Sheet (welding)
	L115		0.7	1.0	0.8		TD	240	295	8	Plate (welding)
	L160	1.6			2.5	5.6	TF over-aged	420	485	8	Extruded section

Figure 8.6 Examples of heat-treatable wrought aluminium alloys

a suffix, which describes the condition and includes strain hardening (as a result of cold working) as well as for heat treatment. The temper designations are: M, as manufactured; O, annealed to obtain the lowest strength condition; H1–H8 strain hardened by cold working after annealing. The designations are in ascending strength order: TB, solution treated and naturally aged; TD, solution treated, cold worked and naturally aged; TE, cooled from forming process and precipitation treated; TF, solution treated and precipitation treated; TH, solution treated, cold worked and then precipitation treated.

Wrought aluminium alloys

These alloys can be 'strengthened' either by heat treatment or by work-hardening.

Some heat-treatable alloys become spontaneously harder and stronger after solution treatment while others require further precipitation treatment to develop their optimum properties. Alloys in this group can be considered as those containing copper in the order of 4.5% magnesium and silicon about 1.5%, and those significant amounts of copper together with magnesium and silicon. In addition, special alloys developed for the aerospace industry are also available, the most significant and the strongest containing 6% zinc. Figure 8.6 lists typical heat-treatable wrought alloys.

Most of the strain-hardening wrought alloys have a solid solution structure whilst others have a solid solution structure with small amounts of a second phase present. This is demonstrated by figure 8.7 which shows that only a small

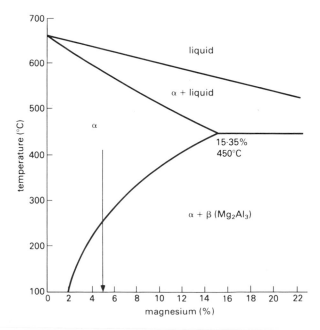

Figure 8.7 Aluminium-rich part of the magnesium–aluminium thermal equilibrium diagram.

Type	Designation	Basic composition Mn (%)	Si (%)	Mg (%)	Treatment	0.2% proof (N/mm²)	UTS (N/mm²)	Elongation (%)	Applications
BS1470 to BS1475	4047A		12.0		M	—	—	—	Wire
	4043A		5.2		M	—	—	—	Wire
	3103	1.2			H8	—	175	4	Plate
	5251	0.3		2.0	H4	175	225	5	Drawn tube
	5154A	0.3		3.5	M	100	215	16	Forgings
	5056A	0.3		5.0	H2	—	280	—	Rivet stock
	5556A	0.8		5.2	M	—	—	—	Wire
	5083	0.7		4.5	O	125	275	11	Extruded section
Aerospace 'L' series	4L44		2.0		O	60	170	16	Extruded section
	3L60	1.1			H2	120	145	9	Sheet
	2L81			2.0	H6	175	250	4	Sheet

Figure 8.8 Examples of strain-hardening wrought aluminium alloys

amount of the second phase is present in this particular alloy. Due to the solid-solution structure these alloys are soft, possess very good corrosion resistance and lend themselves to strengthening by cold working—at the expense of ductility.

Alloys in this group can be considered as those containing up to 2% alloying element and those containing up to 5% magnesium with small amounts of manganese. Figure 8.8 lists typical strain-hardening wrought alloys.

Cast aluminium alloys

Where good casting properties are required a 'fluid' alloy containing in excess of 5% silicon is used. The more silicon added, the more fluid the alloy. In addition the solidification range is reduced, as shown in the aluminium–silicon equilibrium diagram in figure 8.9.

Up to the eutectic composition of 11.7% silicon the structure will be of primary α-solid solution in a fine eutectic matrix. However, above the eutectic composition, the structure possesses coarse primary silicon in an α-solid-solution

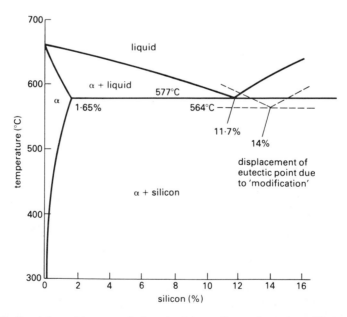

Figure 8.9 Aluminium-rich part of the aluminium–silicon thermal equilibrium diagram, showing the effect of modification on the eutectic.

Figure 8.10(a) Microstructure of an 'as-cast' aluminium–12% silicon alloy. The coarse brittle structure consists of dark-coloured silicon eutectic in an α-solid solution matrix.

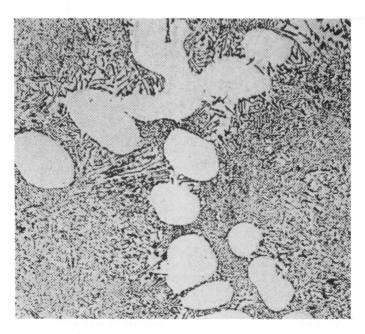

Figure 8.10(b) Microstructure of the same alloy as in Figure 8.10(a) which has been 'modified'. The refined structure now consists of light-coloured tough primary α-solid solution in a eutectic matrix.

matrix—an undesirable structure for engineering purposes (figure 8.10(a)). To overcome this problem for very fluid alloys with a silicon content in excess of eutectic composition, the alloys are inoculated just prior to casting which reduces the eutectic temperature and increases the eutectic composition as indicated in figure 8.9. A 'modified' aluminium–silicon alloy will now have a refined hypoeutectic structure with high fluidity and a reduced melting temperature (figure 8.10(b)).

All casting alloys are suitable for sand casting, most of them being suitable for gravity die-casting. However, only a few are suitable for pressure die-casting, which is restricted to the cold-chamber method due to the tendency of molten aluminium to pick up iron from the cylinder of the casting machine.

Alloys in this group may be strengthened either by alloying or by heat-treatment.

As-cast aluminium alloys

This group of alloys does not respond to heat treatment and rely on alloying for strength where necessary. The group consists of alloys with between 4.5% and 6.0% or between 10% and 13% silicon; 25% silicon and up to 5.0% copper or between 5% and 9% silicon and between 3% and 5% copper; and between 3% and 6% magnesium. Figure 8.11 lists typical as-cast aluminium alloys.

Aluminium alloys which respond to heat treatment contain silicon together with copper. They can be either of high silicon low copper or low silicon high copper. Typical examples are shown in figure 8.12.

8.6   Pure magnesium

Magnesium was first purified by the English scientist Davy as long ago as 1808, in fact before aluminium. Technical and economic problems retarded its development and it was not until 1923 that a satisfactory extraction process was identified. Extraction was uneconomic prior to the development of industrial electrolysis and the demand for such a light-weight metal only came with the development of the aircraft industry.

Magnesium is the lightest of engineering metals and has similar characteristics to aluminium in that its melting temperature is similar and it has an affinity for oxygen. With magnesium, it burns with intense heat. However, although the surface oxide provides sufficient protection against corrosion in a dry atmosphere, its resistance in humid conditions is low. Unlike the more common metals, magnesium has a CPH lattice structure which does not allow slip to take place readily. This restricts the elastic and cold-working properties but provides a relatively high tensile strength—110 N/mm^2 in a cast condition which can be increased by cold working to 180 N/mm^2. Together with the low relative density of 1.7, magnesium provides a high strength: weight ratio.

Magnesium also presents foundry problems. Due to its affinity for oxygen it is necessary to melt the metal under a flux and protect the metal when casting with

Light alloys 121

Type	Designation	Cu (%)	Mn (%)	Si (%)	Mg (%)	Zn (%)	Treatment	Sand cast 0.2% proof (N/mm²)	Sand cast UTS (N/mm²)	Sand cast Elongation (%)	Chill cast 0.2% proof (N/mm²)	Chill cast UTS (N/mm²)	Chill cast Elongation (%)	Applications
General purpose to BS1490	LM2	1.6		10.5		2.0	M	—	—	—	—	150	—	Thin-walled die-casting
	LM6			11.5			M	—	160	5	—	190	7	High corrosion resistance
	LM20	0.4		11.5			M	—	—	—	—	190	5	Lower corrosion resistance than 6
	LM24	3.5		8.5		3.0	M	—	—	—	—	180	1.5	Similar LM6 but die-cast
	LM27	2.0	0.4	7.0			M	—	140	1	—	160	2	Sand and die wide range
Special purpose to BS1490	LM5		0.5		4.5		M	—	140	3	—	170	5	Marine use
	LM12	10.0		2.5	0.3		M	—	—	—	—	170	—	Hydraulic use
	LM18			5.2			M	—	120	3	—	140	4	Cooking application
	LM21	4.0	0.4	6.0	0.2	2.0	M	—	150	1	—	170	1	Good fluid and mechanical property
	LM30	3.0		17.0	0.5		M	—	—	—	—	150	—	Low coefficient of expansion

Figure 8.11 Typical 'as-cast' aluminium alloys

Type	Designation	Basic composition					Treatment	Properties						Applications
								Sand cast			Chill cast			
		Cu (%)	Mn (%)	Si (%)	Mg (%)	Zn (%)		0.2% proof (N/mm²)	UTS (N/mm²)	Elongation (%)	0.2% proof (N/mm²)	UTS (N/mm²)	Elongation (%)	
General purpose to BS1490	LM4	3.0	0.4	5.0			TF	—	230	—	—	280	—	Most used versatile
	LM25			7.0	0.4	1.0	TB7	—	160	2.5	—	230	5	High mechanical properties
Special purpose to BS1490	LM9		0.5	11.5	0.4		TE	—	170	1.5	—	230	2	Low-pressure casting
	LM10			10.5			TB	—	280	8	—	310	12	High strength good impact
	LM13	1.1		11.5	1.2		TF7	—	140	—	—	200	—	High temperature use, low expansion
	LM16	1.2		5.0	0.5		TF	—	230	—	—	280	—	High strength up to 200°C
	LM22	3.3	0.4	5.0			TB	—	—	—	—	245	8	Heavy-duty road vehicles

LM26	3.0	9.5	1.0	1.0	TE	—	—	—	210	—	Internal combustion engine pistons
LM28	1.5	18.5	1.2		TF	—	120	—	190	—	High hardness low coefficient of expansion
LM29	1.1	23.5	1.1		TE	—	120	—	190	—	Similar to 28 better coefficient of expansion
Aerospace 'L' series											
4L35	4.0	1.4			TB	210	220	—	230	280	—	Age-hardening, high strength
3L51	2.4	2.1	0.1		TE	125	160	2	140	200	3	Good general purpose alloy, high rigidity
3L78	1.2	5.0	0.5		TF	220	250	—	250	300	—	Pressure-tight at high temperature
2L99		7.0	0.3		TF	185	230	2	200	280	5	High proof at high temperature

Figure 8.12 Typical heat-treatable casting aluminium alloys

flowers of sulphur which burns off, thus consuming the oxygen in the surrounding air.

As with most pure metals, magnesium is not strong enough for structural applications without alloying. In some cases alloying can increase the ultimate strength to a value in excess of 300 N/mm^2.

8.7   Elements alloyed with magnesium and their effects

Alloys of magnesium first became available in the early 1920s. Since then development has taken place to meet the demands of the aircraft industry and more recently for the nuclear industry as a fuel-element canning material. Magnesium has a low neutron absorption cross-section.

As with aluminium, the mechanical properties of a pure metal can be improved by alloying with elements that can form a solid solution with it. Alternatively improvement can be gained from elements which form strengthening compounds or those which cause the alloy to respond to strengthening heat treatment. Since magnesium has a CPH lattice structure only a limited number of elements will form a solid solution and in these cases only small amounts. Such elements are aluminium, manganese, zinc, thorium and silver. This is illustrated in the thermal equilibrium diagrams in figures 8.13 to 8.17.

In can be seen that with aluminium, zinc and thorium the diagrams are of the form shown in figure 8.1, in which the solid solubility increases with temperature, a situation which facilitates precipitation treatment as described in Section 8.1.

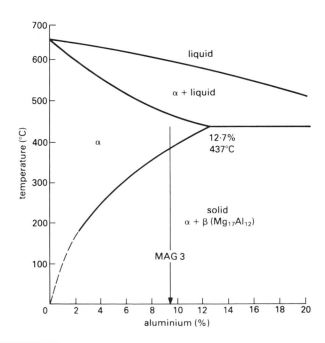

Figure 8.13 Magnesium-rich part of the magnesium–aluminium thermal equilibrium diagram.

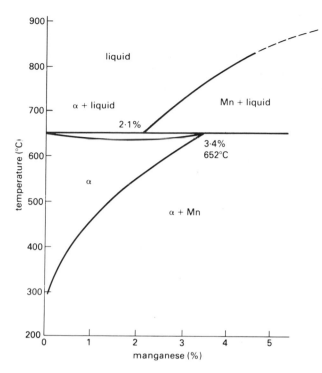

Figure 8.14 Magnesium-rich part of the magnesium-manganese thermal equilibrium diagram.

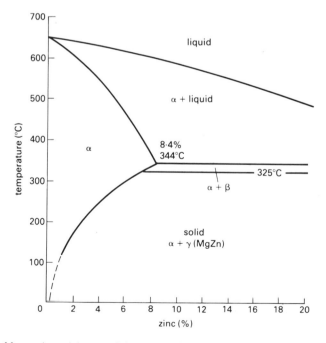

Figure 8.15 Magnesium-rich part of the magnesium–zinc thermal equilibrium diagram.

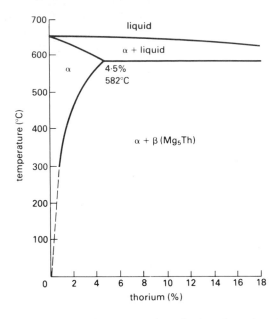

Figure 8.16 Magnesium-rich part of the magnesium–thorium thermal equilibrium diagram.

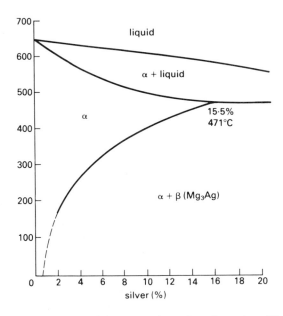

Figure 8.17 Magnesium-rich part of the magnesium–silver thermal equilibrium diagram.

Small amounts of *rare earth metals* have a similar effect but in this case silver is used to accelerate the ageing.

Manganese is used in magnesium alloys to improve corrosion resistance whereas zirconium is used as a grain refiner thus improving strength.

In comparing the aluminium, zinc and thorium equilibrium diagrams it will be seen that there is a significant difference in the eutectic temperatures. Hence, whereas aluminium is used as a general purpose hardening agent thorium is used for higher temperature applications, improving creep resistance and zinc is used in low temperature applications.

8.8   Heat treatment of magnesium alloys

Like aluminium alloys, some magnesium alloys respond to strengthening heat treatment. Such alloys possess limited solid solubility which increases with temperature. Figure 8.13 illustrates the magnesium-rich part of the magnesium–aluminium thermal equilibrium diagram and the precipitate as β-compound. Again, the amount of alloying element, aluminium, must be less than the maximum which can be contained by the solid solution, i.e. 12.7% at 437°C.

Consider a magnesium alloy containing 9.5% aluminium, say to BS 2970 MAG 3. This particular alloy also contains small amounts of zinc and manganese which necessitates a two-stage solution treatment due to grain boundary composition. When the alloy is heated at 385°C for 8 h followed by heating at 415°C for 16 h all the aluminium will be taken into solid solution. If the alloy is then cooled in air there will be insufficient time for any precipitation and a supersaturated solid solution will result. This *solution treatment* produces a soft, ductile material.

In order to restore equilibrium conditions the alloy must be reheated to 200°C for 10 h during which time β-precipitation will take place. This *precipitation treatment* will increase the strength and hardness of the alloy. The changes in microstructure can be seen in figures 8.18–8.20

As stated in Section 8.7, other elements which may be alloyed with magnesium to produce a heat-treatable alloy include *rare earth metals*. Manufacturers have developed high-strength, creep-resisting alloys which contain neodymium(Nd)-rich rare earth metals to which silver has been introduced to speed up the ageing. These alloys contain either 2.0 or 2.5% Nd-rich rare earth metals together with 2.5% silver and 0.6% zirconium. The manufacturers recognise the effect of silver on the cost of the alloys, particularly when the cost of silver escalated dramatically recently, and efforts were made to find some means of reducing the silver content without impairing the properties. Magnesium Elektron Ltd. have succeeded in producing an Nd–rare earth magnesium alloy in which the silver content has been reduced by 32% by controlled additions of copper. This alloy contains only 1.3–1.7% silver, compared with 2.0–3.0%, together with the introduction of 0.05–0.1% copper.

8.9   Magnesium alloys

Magnesium alloys can be grouped in a similar manner to the grouping of aluminium alloys. The identification system devised by the British Standards Institute is also similar. The current nomenclature is briefly as follows.

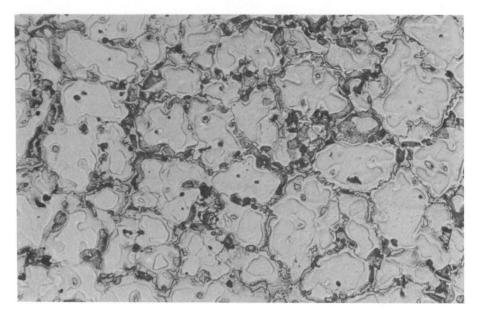

Figure 8.18 Microstructure of a magnesium–9.5% aluminium casting alloy in the 'as-cast' condition (BS2970 MAG 3 M). The cored nature of the grains are revealed together with the compound $Mg_{17} A_{12}$ and lamellar eutectoid precipitate at the grain boundaries. (Courtesy of Magnesium Elektron Ltd).

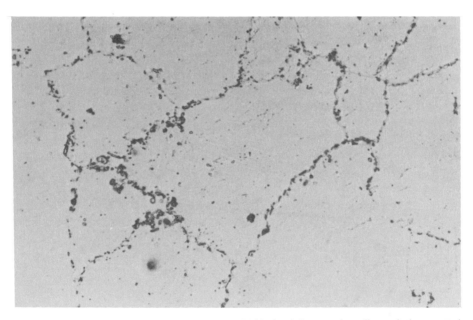

Figure 8.19 Microstructure of a magnesium–9.5% aluminium casting alloy solution treated (BS2970 MAG 3 TB). The grain boundaries of the essentially single-phase structure are revealed by the precipitate $Mg_{17} A_{12}$. (Courtesy of Magnesium Elektron Ltd).

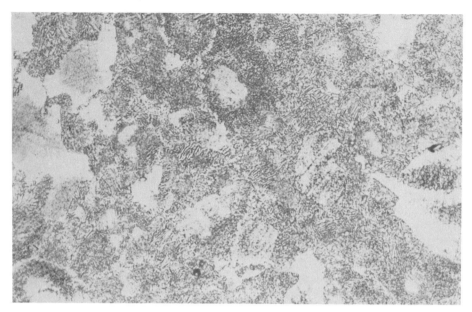

Figure 8.20 Microstructure of a magnesium–9.5% aluminium casting alloy solution treated and precipitation treated (BS2970 MAG 3 TF). The structure reveals large area of the grains with the lamellar eutectoid precipitate which grows progressively from the grain boundaries. (Courtesy of Magnesium Elektron Ltd).

All alloys carry the prefix MAG identifying them as magnesium alloys. Where the alloy is used for wrought products the letters MAG are followed by a letter, separated by a hyphen, indicating the form of supply: i.e. S, plate, sheet and strip (BS 3370); F, forgings and forging stock (BS 3372); E, extruded bars, sections and tubes (BS 3373).

In the case of casting alloys (BS 2970) there is no symbol to indicate form. The identification is followed by a number, separated by a hyphen, indicating the particular alloy in the series. In addition the identification carries a final letter suffix, separated by a hyphen, which indicates the condition of the material. The form symbols are as follows: O, annealed to the softest condition; M, as-manufactured, i.e. as-cast, as-rolled, as-extruded etc.; TS, stress relieved only; TE, precipitation treated only; TB, solution treated only; TF, solution treated and precipitation treated. A tube to BS 3373, as extruded would be identified as MAG-E-121-M

In some cases a magnesium alloy may be identified by a *compositional designation*. In this designation the basic metal symbol Mg appears first, separated by a hyphen from the essential alloying elements in diminishing order of content. Where the essential element is less than 1% the element symbols have no following number. Others are followed by a number indicating the percentage of the element to the nearest 0.5%, i.e. MAG-E-121-M (Mg-Al6Zn1Mn). Where the number indicating the alloy in the series is a single digit it indicates a casting alloy whereas three digits indicate a wrought alloy.

Wrought magnesium alloys

With the exception of one alloy listed in BSS 3370, 3372 and 3373, all the wrought alloys are specified in the 'as-manufactured' or 'full-annealed' condition. Hence wrought magnesium alloys rely on alloying for strength. Since magnesium has a

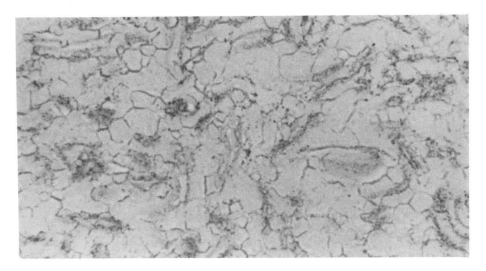

Figure 8.21 Microstructure of a magnesium–3% ZrZn wrought alloy (BS3373 MAG-E-151) which has been recrystallised. It reveals an extremely fine grain size due to zirconium refinement. (Courtesy of Magnesium Elektron Ltd).

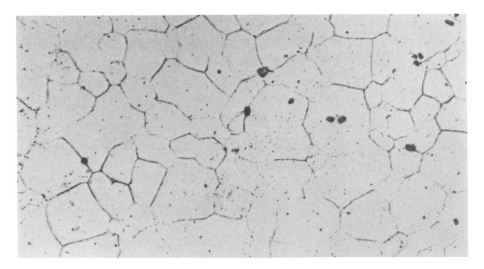

Figure 8.22 Microstructure of a magnesium–6% AlZnMn wrought alloy (BS3373 MAG-E-121) which has been recrystallised. It reveals equi-axed grains of solid solution together with manganese inclusions. (Courtesy of Magnesium Elektron Ltd).

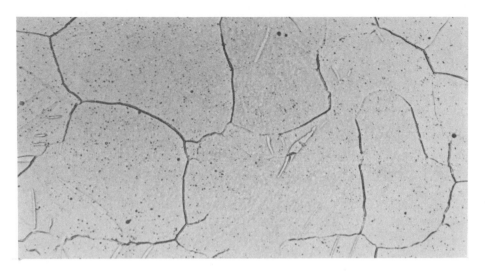

Figure 8.23 Microstructure of a magnesium–manganese wrought alloy (BS3373 MAG-E-101) which has been recrystallised. It reveals an equi-axed grain structure in which there is evidence of a manganese precipitate. (Courtesy of Magnesium Elektron Ltd).

CPH lattice structure—a structure which does not readily slip—the alloys cannot be cold worked. They can, however be shaped by hot working which is conducted at temperatures above that of recrystallisation. As the metal cools from a recrystallised structure the resulting material will consist of relatively small equi-axed, stress-free grains.

Wrought magnesium alloys to British Standards can be classified according to composition, with alloys containing up to 1.5% manganese, up to 6% aluminium with zinc and manganese, up to 2% zinc with manganese and up to 6% zinc with zirconium. Alloys in the last classification with high zinc content are the only wrought alloys which are heat treated. Illustrations of the microstructures of typical wrought alloys are shown in figures 8.21–8.23 and examples of wrought alloy compositions and properties are shown in figure 8.24.

Cast magnesium alloys

All the casting alloys of magnesium can be strengthened by heat treatment if necessary, as indicated in BS 2970 and the *Aerospace 'L' Series* specifications. As with wrought alloys, they can be classified according to composition. The groups include alloys containing between 8 and 10% aluminium with zinc and manganese, alloys containing zinc, zirconium and thorium, which are creep-resisting, alloys containing rare earth metals, zirconium and silver, alloys containing up to 4.5% zinc with thorium and those alloys which contain zinc, rare earth metals and zirconium. Examples of typical casting alloys are shown in figure 8.25.

Type	British Standard designation	Compositional designation	Basic composition				Treatment	Properties			Form
			Al (%)	Zn (%)	Mn (%)	Zr (%)		0.2% proof (N/mm^2)	UTS (N/mm^2)	Elongation (%)	
BS3370, BS3372, BS3373	MAG-E-101	Mg-Mn 1.5			1.5		M	120	230	4	Extrusion
	MAG-S-111	Mg-Al3Zn 1 Mn	3.0	1.0	0.3		O	160	245	10	Sheet
	MAG-F-121	Mg-Al6Zn1Mn	6.0	1.0	0.3		M	160	270	7	Forging
	MAG-S-131	Mg-Zn2Mn1		2.0	1.0		O	120	265	10	Sheet
	MAG-E-141	Mg-Zn1Zr		1.0		0.6	M	185	260	8	Extrusion
	MAG-F-151	Mg-Zn3Zr		3.0		0.6	M	180	270	7	Forging
	MAG-E-161	Mg-Zn6Zr		6.0		0.6	TE	230	315	8	Extrusion
Aerospace 'L' series	2L509	—		1.2		0.6	M	170	250	5	Tube (welding)
	L512	—	6.0	1.0	3.0		M	180	270	8	Bar stock
	L514	—	3.0			0.6	M	205	290	8	Forging

Figure 8.24 Typical wrought magnesium alloys

Type	British Standard designation	Compositional designation	Al (%)	Zn (%)	Mn (%)	RE (%)	Zr (%)	Th (%)	Condition	UTS (N/mm²) Sand	UTS (N/mm²) Chill	0.2% proof (N/mm²) Sand	0.2% proof (N/mm²) Chill	Elongation (%) Sand	Elongation (%) Chill	Uses
General purpose to BS2970	MAG1	Mg-Al8ZnMn	8.0	0.5	0.3	—	—	—	M	140	185	85	85	2	4	Good mechanical properties. Automobile road wheels
	MAG3	Mg-Al10ZnMn	10.0	0.5	0.3	—	—	—	M	125	170	95	100	—	2	Pressure tight. Used for engine manifold covers
	MAG4	Mg-Zn4.5Zr		4.5			0.7	—	TE	230	245	145	145	5	7	A structural alloy with high proof stress
	MAG7	Mg-Al8.5Zn1Mn	8.5	1.0	0.5			—	TF	185	215	110	110	—	2	A good general-purpose alloy
Special purpose to BS2970	MAG2	Mg-Al8ZnMn	8.0	0.6	0.4	—	—	—	M	140	185	85	85	2	4	Used for instrument casings
	MAG5	Mg-Zn4REZr		4.0		1.2	0.7	—	TE	200	215	135	135	3	4	A high-strength, pressure-tight alloy
	MAG6	Mg-RE3ZnZr		1.0		3.0	0.7	—	TE	140	155	95	110	3	3	A high-temperature, pressure-tight alloy
	MAG8	Mg-Th3ZnZr		2.0			0.7	3.0	TE	185	185	85	85	5	5	Creep-resistant up to 350°C
	MAG9	Mg-Zn5.5Th2Zr		5.5			0.7	2.0	TE	225	225	155	155	5	5	Used for high-duty structures
Aerospace 'L' series	3L122	—	8.0	0.5	0.3	—	—	—	TB	200	230	80	80	7	10	Similar to MAG1
	3L125	—	10.0	0.5	0.3	—	—	—	TF	200	120	120	120	—	2	Similar to MAG3
	2L126	—		2.3		3.0	0.6	—	TE	140	155	95	100	3	3	Similar to MAG6
	2L127	—		4.5			0.7	—	TE	230	245	145	145	5	7	Similar to MAG4
	2L128	—		4.5		1.2	0.7	—	TE	200	215	135	135	3	4	Similar to MAG5

Figure 8.25 Typical commercial magnesium casting alloys.

Chapter 9

Copper and zinc alloys

9.1　Pure copper

Copper was the first metal used by man and it is a considered opinion that bronze was accidentally produced in Cornwall by the ancient Britons when smelting ore containing both tin and copper. During the eighteenth century most of the world's ore was mined in Cornwall or Wales and it is no surprise to learn that Swansea became the world's smelting centre. When other larger deposits were found in places such as South America, Africa and the USSR, Britain ceased to be the centre of the copper industry.

Copper, of relative density 8.9, possesses many attributes which make it a very useful metal in both its unalloyed and alloyed condition. Since pure copper has a FCC lattice structure, this influences its mechanical properties, its tensile strength and hardness being relatively low with a high measure of ductility. Cast copper has an ultimate strength of up to 150 N/mm^2 which can be significantly increased to 400 N/mm^2 by cold working. Hardness increases correspondingly from 45 to 90 Brinell hardness units.

The most significant physical property is its very high electrical conductivity, which is second only to silver. Hence most of the copper produced is used in electrical applications. Other relevant properties that copper possesses are good thermal conductivity and corrosion resistance, making it a suitable material for water-heating systems.

9.2　Commercial grades of pure copper

There are two methods of refining copper, namely fire refining and electrolytic, the method by which the copper is refined dictating the application. In the fire-refining method the smelted copper is remelted in a furnace where most of the impurities oxidise and are lost as slag. However, electrolytic refining produces high-purity copper by electrolysis. The 'cathode' copper so produced is 99.97% pure.

There are four principle commercial grades of copper.

Electrolytically refined, oxygen-free, high-conductivity copper (OFHC)

This is 'cathode' copper, the purest form of copper. It is 99.97% pure and hence has the highest degree of electrical conductivity. It should be noted that any slight decrease in purity will result in a significant decrease in electrical conductivity.

Tough-pitch copper

This is copper produced by fire-refining but contains small amounts of impurities in the form of copper oxide. The oxide is present as small globules, which, although it does not greatly affect the mechanical properties, has a significant effect on the electrical conductivity and the thermal properties. Its presence also affects flame-welding capabilities. In these circumstances the oxides react with the flame and liberate gases which tend to form blow-holes—an undesirable situation.

Deoxidised copper

This is tough-pitch copper treated with a small amount of phosphorus just before casting to remove the copper oxide globules. Although the residual phosphorus does not impair welding, even amounts as small as 0.04% reduce the electrical conductivity by as much as 25%.

Arsenical copper

When small amounts of arsenic are added to copper it goes into solid solution. In commercial copper about 0.5% arsenic is added to increase the corrosion resistance and strength. Although the resulting electrical conductivity is poor, the softening temperature is increased from 200 to 500°C, thus making the material suitable for boiler tubes.

9.3   Elements alloyed with copper and their effects

The principle elements alloyed with copper to improve the properties are zinc, tin, aluminium and nickel. When alloyed individually they produce the widely used brasses, bronzes, cupro-nickels and nickel-silvers.

When copper is alloyed with *zinc* the resulting alloy is known as brass. Due to the similar atomic size of the elements and the fact that under these conditions a CPH lattice-structured zinc will 'fit' reasonably well within the FCC lattice structure of the copper, the alloy gives rise to a wide range of solid-solution phase. This is shown in the thermal equilibrium diagram figure 9.1(a). Since the presence of zinc atoms within the copper lattice gives rise to some distortion there will be a progressive increase in strength as the zinc content increases, reaching a maximum at 42% zinc. However, an unusual feature is that the ductility of the alloy also increases with dissolved zinc content, reaching a maximum value with 30% zinc, which then decreases with further additions of zinc. At 37% zinc the alloy has a ductility value equal to that of pure copper. Such variations in properties are illustrated in figure 9.1(b).

An alloy of copper and *tin*, produce tin bronze. There is a distinct difference in both the relative atomic mass of the elements and in the type of lattice structure they assume. This arrangement gives rise to distortion of the copper lattice at quite low percentages of tin with a very limited solid solution phase at room

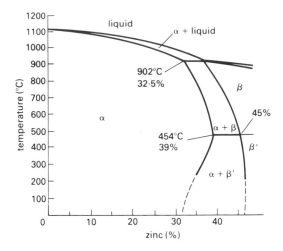

Figure 9.1(a) Copper–zinc thermal equilibrium diagram.

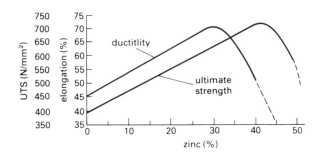

Figure 9.1(b) Relationship between zinc content and mechanical properties.

temperature. This is illustrated in the thermal equilibrium diagram shown in figure 9.2. Due to the limited solid-solution range there is limited ductility. The main advantages in alloying tin with copper are in the physical properties, particularly its good corrosion resistance and good wear resistance. Tin also imparts good casting properties.

When *aluminium* is alloyed with copper to produce aluminium bronze, the relative atomic mass of the elements and the type of lattice structure they form is more amenable to the formation of a solid solution. Although the relative atomic mass of aluminium is half that of copper, both elements are of a FCC lattice structure. The thermal equilibrium diagram, shown in figure 9.3, indicates that up to 9.5% aluminium can be contained in the solid solution. However, unlike other thermal equilibrium diagrams previously considered, solubility does not increase with temperature. Solubility, in fact, remains constant up to 565°C and then decreases with any further rise. Hence, alloys with, say, 7% aluminium content will be capable of being strengthened by cold working after annealing but alloys in excess of this content can only be hot-worked or cast. In general

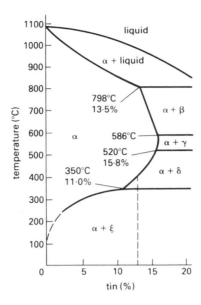

Figure 9.2 Copper–tin thermal equilibrium diagram.

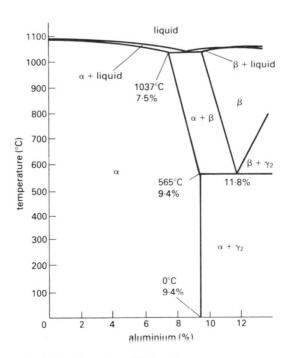

Figure 9.3 Copper–aluminium thermal equilibrium diagram.

copper–aluminium alloys have good corrosion resistance due to the alumina film produced on the surface and are capable of retaining their properties at reasonably high temperatures.

Cupro-nickels are produced when copper is alloyed with *nickel*. In this alloy both elements form a FCC structure and have an almost identical relative atomic mass. This gives rise to a near-perfect substitutional solid solution in which there is complete solubility in all proportions, as illustrated in the thermal equilibrium diagram figure 9.4. Since no second phase can be present all copper–nickel alloys can be cold-worked, strength increasing with nickel content but ductility remaining fairly constant. The main advantage of such alloys is their very high corrosion resistance.

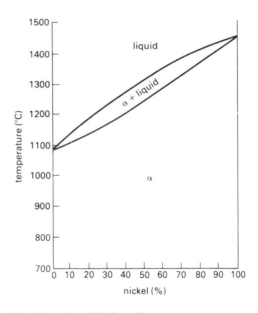

Figure 9.4 Copper–nickel thermal equilibrium diagram.

If zinc is added to a copper–nickel alloy at the expense of copper, a nickel–silver alloy is produced. This addition of zinc will introduce a second phase which will facilitate cold working.

Other elements may be added in small quantities to the main copper alloy systems described previously. Such elements are tin, zinc, lead, nickel, iron, aluminium and manganese.

Up to 2% tin may be added to copper–zinc alloys to improve their resistance to corrosion.

To improve the fluid properties in casting copper–aluminium, alloys up to 0.5% zinc is added. The same effect is derived by adding up to 8% to copper–tin alloys but in this system it also acts as a deoxidiser. Since zinc is cheaper than tin any significant addition of zinc at the expense of tin will produce a cheaper alloy.

Lead is a common additional alloying element, in all cases improving machinability. In copper–zinc alloys, up to 4% is added, whereas in copper–nickel–zinc alloys the amount is up to 2%. Up to 5% lead is added in copper–tin alloys which has a further effect of improving castability. Where copper–tin alloys are used for bearings, up to 20% may be added to improve the thermal conductivity.

Nickel may be added to brasses and bronzes to improve corrosion resistance and mechanical properties at high temperatures. Up to 1% may be found in copper–zinc alloys and up to 5%, in conjunction with iron and manganese, in copper–aluminium bronzes. In copper–tin bronzes up to 5.5% may be added in conjunction with zinc to improve the mechanical properties and in these alloys, where the nickel content is at the limit, it may be precipitation hardened to further improve strength.

Iron, in conjunction with manganese, nickel and aluminium is added to copper–zinc alloys up to 4.5% to improve their strength. Similar amounts, in conjunction with nickel, are used in copper–aluminium alloys to the same effect through grain refinement. However, in copper–nickel alloys, up to 1.5% induces age-hardening, increasing strength and corrosion resistance.

Aluminium is an additional element found in copper–zinc alloys. Up to 4.5%, in conjunction with manganese, nickel and iron is used to improve strength in the *high-tensile* alloys.

Manganese is widely used to improve strength and toughness. In copper–zinc alloys, up to 4%, in conjunction with iron and aluminium is used to improve strength in high-tensile alloys, whereas up to 1.5%, in conjunction with iron, is used in copper–aluminium alloys for the same purpose. In both types of

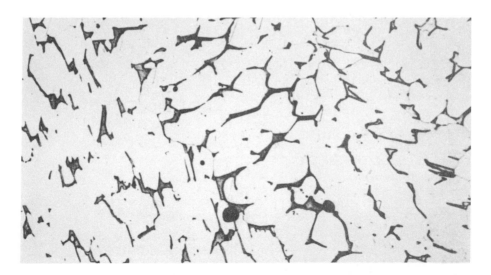

Figure 9.5 Microstructure of a brass containing 37% zinc in the 'as-cast' condition. The structure consists mainly of the lighter α-phase solid solution with relatively small, darker, angular crystals of the hard β' phase at the grain boundaries and within the α grains.

Type	BS number	Designation	Basic composition							Condition	Mechanical properties				Uses
			Zn (%)	Sn (%)	Pb (%)	Ni (%)	Fe (%)	Al (%)	Mn (%)		UTS (N/mm²)	0.2% proof (N/mm²)	Elongation (%)	Hardness (H$_B$)	
Forgings	2872	CZ109	40	—	—	—	—	—	—	M	310	—	25	—	Hot-working 'Muntz metal'
		CZ112	38	1.2	—	—	—	—	—	M	340	—	15	—	Naval brass, good corrosion resistance
		CZ114	37	0.5	0.75	—	0.75	1.5	1.0	M	460	195	15	—	High tensile brass, 'manganese bronze', marine applications
		CZ121 3Pb	40	—	3.0	—	—	—	—	M	310	—	15	—	Free-cutting brass, for high-speed machining
		CZ121 4Pb	39	—	4.0	—	—	—	—	M	310	—	15	—	Free-cutting brass, for high-speed machining
		CZ123	40	—	0.5	—	—	—	—	M	310	—	25	—	Hot working
Extrusion	2874	CZ106	30	—	—	—	—	—	—	M	340	—	28	—	'Cartridge brass', maximum ductility, deep drawing
		CZ124	35	—	3.0	—	—	—	—	H	480	220	3	—	'Standard brass', cold working, free cutting
Sheet & Strip	2870	CZ101	10	—	—	—	—	—	—	H	350	—	3	110	'Gilding metal', architectural metal work
		CZ107	35	—	—	—	—	—	—	EH	525	—	—	165	'Standard brass', general purpose cold-working alloy, lead free

Castings									Sand	Chill	Sand	Chill	Sand	Chill	Sand	Chill	
1400A	SCB3	30	1.0	2.0	1.0			M	200	—	90	—	20	—	55	—	General purpose sand-casting alloy
	SCB6	15	0.5					M	180	—	90	—	30	—	55	—	General purpose sand-casting alloy
	DCB3	38	1.0	1.5				M	—	320	—	105	—	28	—	65	General purpose gravity die-casting alloy
	PCB1	40	0.5	1.5				M	—	330	—	105	—	33	—	65	General purpose pressure die-casting alloy
1400B	HTB1	43	1.0	0.5	1.0	1.5	3.0	M	520	535	225	245	26	26	125	—	High-tensile casting alloy
	HTB3	43	0.2	0.2	1.0	2.5	4.5	4.0	M	775	—	435	—	14	—	190	—	High-tensile sand-casting alloy
1400C	SCB4	37	1.5	0.5				M	280	—	90	—	30	—	62	—	Sand-casting alloy, limited availability

Figure 9.6 Typical commercial brasses

copper–nickel alloys, the cupro-nickels and the nickel-silvers, manganese is residual from deoxidation, thus increasing toughness.

Phosphorus is added to copper–tin alloys producing what is known as phosphor-bronze. Up to 0.4% is used to improve strength and hardness at the expense of ductility and also to improve corrosion resistance.

9.4   Copper alloys and their applications

Copper alloys can be classified into groups according to their major alloying element, namely brasses, tin bronzes, aluminium bronzes, cupro-nickels and nickel-silvers. Within each group a range of alloys are available depending on the specific composition, which in some cases defines the alloy as a wrought or a casting alloy. With one exception, all the alloys referred to achieve maximum properties through alloying or cold working and not through heat treatment.

Brasses

These are alloys of copper and up to 45% zinc. They may also contain small quantities of lead, tin, nickel, iron, aluminium and manganese. The copper–zinc thermal equilibrium diagram (figure 9.1(a)) shows that copper will dissolve up to 32.5% zinc forming an α-phase solid solution at the solidus temperature of 902°C, the zinc content increasing to 39% as the temperature falls to 454°C. With further slow cooling under equilibrium conditions the zinc content reduces to 35.2% at 250°C. However, under normal industrial cooling rates it is found that up to 39% zinc can be dissolved in the α-phase solid solution at room temperature. Since the solubility of the solid solution does not increase with temperature, brasses will not respond to precipitation-hardening treatment. However, brasses with up to 39% zinc content can be strengthened by cold working after annealing. Figure 9.5 shows the microstructure of a brass containing 37% zinc in the as-cast condition. The dark areas are the β'-phase which is present at the boundaries of the α-solid-solution crystals due to industrial cooling rates. Annealing will refine the structure, including the β'-phase thus facilitating cold working.

Such cold-working brasses are known as α-brasses and in order to prevent any loss in ductility, such alloys should be of a high degree of purity. In order to increase the amount of cold working, interstage annealing can be carried out at 600°C. Unfortunately, heavily worked hard-drawn α-brasses are subject to season cracking, resulting from intercrystalline corrosion developing. This can be avoided by low-temperature, stress-relief annealing at 250°C after working.

From the equilibrium diagram it can be seen that any alloy with a zinc content in excess of 39% will contain a second phase, β'. The presence of the β'-phase limits the cold-working capabilities of the alloy, since it is hard and quite tough at room temperature. Such alloys, therefore, can only be hot-worked, an operation made easier by the fact that, above 454°C, the β'-phase changes its structure to a more plastic β-phase. Following such hot-working at 700°C the β-phase reforms upon cooling producing a refined, granular $\alpha + \beta'$ structure. Brasses within this composition range are known as $\alpha + \beta'$ or hot-working alloys. Examples of typical commercial brasses are shown in figure 9.6.

Tin bronzes

These alloys, as the name suggests, are alloys of copper and up to 12% tin. They may also contain small quantities of phosphorus, zinc, lead and nickel. The copper–tin equilibrium diagram (figure 9.2) indicates that up to 15% tin can be dissolved in copper to form the α-phase solid solution at 520°C and that the solubility will fall with a fall in temperature. In fact, due to the coring and slow rate of diffusion that occurs during casting, equilibrium conditions only prevail if the rate of cooling is extremely slow, such as would never be encountered industrially. Hence, it may be assumed that the structure existing at 400°C will be retained down to room temperature when the alloy is cooled under industrial conditions. Figure 9.7(a) is a micrograph of a tin-bronze containing about 4% tin in the as-cast condition. It illustrates the presence of coring, which is typical of a solid solution cooled under non-equilibrium conditions. The thin black lines are the grain boundaries within which appears dark patterns—the cored structure. This condition can be removed by annealing at about 700°C. Figure 9.7(b) is a micrograph of the same material subjected to this process which makes the grains more uniform in composition. The microstructure is seen as isolated tin-rich concentrations within the grain boundaries of the α-solid solution matrix.

Since the equilibrium diagram can be modified below 400°C, it suggests that structure of an α-phase solid solution can contain up to about 14% tin and that the hard, brittle δ-phase is present in alloys with a higher tin content. Tin bronzes can therefore be classified as wrought α-alloys which can be cold worked and

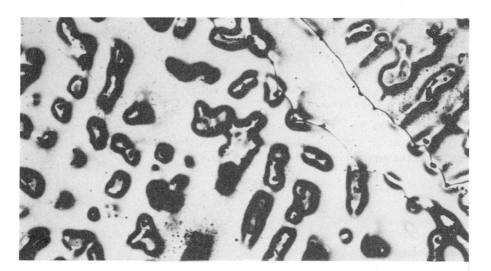

Figure 9.7(a) Microstructure of a 4% tin bronze 'as cast'. The thin black lines represent the grain boundaries if the α-phase solid solution within which are to be seen dark patterns of the tin-rich cored structure.

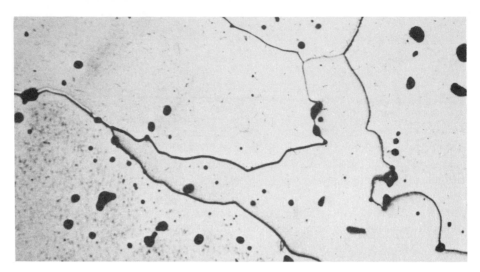

Figure 9.7(b) Microstructure of the same 4% tin bronze after annealing. This reveals only traces of the original cored structure within the grain boundaries of the α-phase solid solution matrix.

$\alpha + \delta$ casting alloys. Wrought alloys contain up to 7% tin although alloys with a higher content can be cold worked following extensive annealing. Casting alloys can contain up to 18% tin but the usual tin content does not exceed 12%.

When between 0.3 and 0.5% phosphorus is added to a tin-bronze the alloy is described as a phosphor-bronze and when between 2 and 8% zinc is added the alloy is described as a gun metal. Examples of typical commercial tin bronzes are shown in figure 9.8.

Aluminium bronze

These bronzes are sometimes described as high-tensile bronzes and are alloys of copper and up to 10% aluminium. They may also contain small quantities of nickel, iron and manganese. From the copper–aluminium equilibrium diagram (figure 9.3) it can be seen that up to 9.4% aluminium will enter into solid solution with copper at room temperature, the amount remaining constant up to 565°C. Above this temperature the amount reduces to 7.5% at the solidus. The diagram also indicates a eutectoid reaction at 565°C and a eutectoid composition of 11.8%. This part of the diagram resembles the steel portion of the iron–iron carbide thermal equilibrium diagram, the α-phase corresponding to ferrite, the β-phase to austenite, the γ_2-phase to cementite and the $\alpha + \gamma_2$ eutectoid to the pearlite eutectoid. Alloys of an aluminium content appropriate to this portion of the diagram can therefore be quenched to produce a structure which resembles martensite. This can then be tempered to produce intermediate properties as with steel. The resulting structures are illustrated in the micrographs in figure 9.9. Figure 9.9(a) is of a copper–12% aluminium cast alloy which has been reheated to 900°C and then slowly cooled. The micrograph reveals a fine lamellar eutectoid

Form	Alloy	British Standard number	Designation	Sn (%)	Al (%)	Zn (%)	Pb (%)	Ni (%)	Fe (%)	P (%)	Mn (%)	Condition	UTS (N/mm²) Sand	UTS Chill	0.2% proof (N/mm²) Sand	0.2% proof Chill	Elongation (%) Sand	Elong Chill	Hardness H_B Sand	Hardness Chill	Uses
Extrusion	Tin bronze	2870	PB101	4.0	—	—	—	—	—	0.3	—	EH	620		580		—		180		Cold-worked phosphor bronze springs
		2874	PB102	5.0	—	—	—	—	—	0.3	—	SH	—		—		—		215		Cold-worked phosphor bronze springs
Forging	Aluminium bronze	2872	CA103	—	9.0	—	—	—	4.0	—	—	M	520		215		20		—		Hot-working alloys, can be heat-treated by quenching and tempering
			CA104	—	10.0	—	—	5.0	5.0	—	—	M	700		400		10		—		
Casting	Tin bronze	1400A	LG2	5.0	—	5.0	5.0	2.0	—	—	—	M	235	240	115	125	20	11	70	87	Leaded gun metal
		1400B	CT1	10.0	—	—	—	—	—	0.15	—	M	270	300	145	165	13	10	80	110	Special-purpose alloy
			LB5	5.0	—	1.0	20.0	2.0	—	0.10	—	M	175	200	80	95	7	8	55	60	Special purpose leaded alloy
		1400C	G1	10.0	—	2.2	1.5	1.0	—	—	—	M	300	270	145	145	20	5	82	110	Gun metal for pump valves etc.
	Aluminium bronze	1400B	AB1	—	9.5	0.5	—	1.0	2.5	—	1.0	M	545	580	185	235	30	30	115	145	Most widely used aluminium bronze

Figure 9.8 Typical commercial tin and aluminium bronzes

Figure 9.9(a)   Micro structure of a copper–12% aluminium cast bronze alloy which has been annealed from 900°C. The micrograph reveals a fine lamellar eutectoid structure comprising light-coloured α-solid solution and dark-coloured hard γ-compound.

Figure 9.9(b)   Microstructure of the same 12% aluminium bronze alloy which has been water-quenched after slow cooling to 530°C. The micrograph shows only a partial transformation to the eutectoid structure, the acicular martensitic-type structure forming upon quenching.

arrangement of light-coloured α-solid solution and dark-coloured γ-compound. When the same alloy is water-quenched after cooling slowly to 530°C, figure 9.9(b) reveals that only a partial transformation to the eutectoid structure has taken place. The micrograph illustrates both the eutectoid structure which formed during the slow cooling together with an acicular structure which formed from the remaining β-phase—similar in appearance to martensite in steel. Quenching from 900°C would therefore produce an entirely acicular structure.

Aluminium bronzes can therefore be classified as α-phase alloys which can be hardened and strengthened by cold working or $\alpha + \gamma_2$ eutectoid alloys which can be hardened and strengthened by heat treatment. Examples of typical commercial aluminium bronzes are shown in figure 9.8.

Cupro-nickels

These alloys contain up to 30% nickel and since nickel forms a solid solution with copper in all proportions they can be readily strengthened by cold working. Alloys with about 20% nickel content are found to be the best for severe cold working, increasing the tensile strength from 340 N/mm² to 540 N/mm². Their exceptional corrosion-resistance is due to the absence of any second phase in the structure and hence there is no possibility of electrolytic corrosion in the presence of an electrolyte. Examples of typical commercial cupro-nickels are shown in figure 9.10.

Nickel-silvers

The description of these alloys tends to be misleading since they do not contain any silver. They are so described due to their silvery appearance. They are, in fact, basically alloys of copper, nickel and zinc, containing between 10 and 18% nickel and between 20 and 30% zinc. The alloys are of a solid solution structure and hence can be cold worked. If, however, hot working is required, impurities must be kept to a minimum. The alloys are often electroplated with silver in which case they are stamped EPNS. Typical commercial nickel-silvers are shown in figure 9.10.

9.5 Pure zinc

Zinc was first identified as such in south-east Asia in the seventeenth century, long after the discovery of brass. It was exported to Europe and England was the first European country to develop its manufacture.

Zinc, of relative density 7.1, has a CPH lattice structure and hence it is relatively strong with low ductility. The structure also limits its forming process to casting. Pure zinc has an ultimate strength of 110 N/mm² and an elongation of 25%.

In order to be a useful engineering material the metal must be 99.99% pure. Any impurities present will be the last to solidify, at the crystal boundaries; this will give rise to eventual intercrystalline corrosion. Since the products of corrosion have a greater volume than the metal they replace it will give rise to swelling of the solid metal with an inevitable increase in brittleness.

Form	British Standard number	Alloy	Designation	Basic composition				Condition	Mechanical properties			Uses
				Ni (%)	Zn (%)	Pb (%)	Mn (%)		UTS (N/mm^2)	Elongation (%)	Hardness (H$_B$)	
Strip	2870	Cupro-nickel	CN104	20	—	—	0.3	O	310	38	—	Best for severe cold working
			CN105	25	—	—	0.2	H	—	—	155	Used for British 'silver' coinage
		Nickel-silver	NS106	18	20	—	0.3	EH	—	—	200	Used for spoting contacts in electrical equipment. Also cutlery
			NS111	10	28	1.5	0.3	1/2H	—	—	150	Leaded nickel-silver, used for Yale-type keys etc.

Figure 9.10 Typical commercial cupro-nickels and nickel-silvers

9.6 Zinc alloys and their applications

Zinc is normally alloyed with aluminium for use as a hot-chamber, die-casting alloy. The zinc–aluminium thermal equilibrium diagram shown in figure 9.11 indicates that only a small amount of aluminium can be taken into solid solution and that a eutectic is formed at 5%. Although the β-phase in the eutectic is unstable, the cooling alloy is subjected to a eutectoid reaction at 275°C to give α + β′. The final structure of an alloy with an aluminium content in excess of 0.35% will be α + β′ which will include the eutectoid structure.

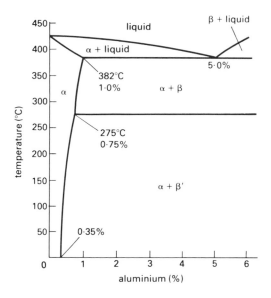

Figure 9.11 Zinc–aluminium thermal equilibrium diagram.

In addition to improving the tensile strength, aluminium reduces the melting temperature thus improving the casting properties. Zinc alloy castings, as with many other cast alloys, are subject to ageing which results from a structural change. This gives rise to a reduction in strength, toughness and hardness and an increase in ductility. During ageing at room temperature much of the 0.35% aluminium in solid solution is rejected in the form of β′ so that the α-phase eventually contains only 0.07% aluminium. This gives rise to slight shrinkage, most of which takes place in the first 5 weeks after casting. This dimensional change can be allowed for and can be accelerated by low-temperature stabilising treatment at 150°C for 3 h after casting.

Zinc is normally alloyed with approximately 4% aluminium which improves the ultimate strength to 285 N/mm² at the expense of ductility, the percentage elongation falling to 10%. However, the ultimate strength can be further improved by adding a small amount of copper; 1% copper increases the ultimate

British Standard Number	Designation	Basic Composition			Condition	Mechanical properties				Shrinkage after 5 weeks natural ageing (mm/m)	Uses
		Al (%)	Cu (%)	Mg (%)		UTS (N/mm^2)	Impact (J)	Elongation (%)	Hardness (H$_B$)		
1004	Alloy 'A'	4.1	—	0.05	M	285	57	10.8	83	−0.65	Wide range of applications in automobile and domestic appliance industries
	Alloy 'B'	4.1	1.0	0.05	M	330	58	6.5	92	−0.83	

Figure 9.12 Typical commercial zinc casting alloys

strength to 330 N/mm² with a further reduction in ductility. Unfortunately the presence of copper increases the amount of shrinkage of the alloy.

Typical commercial zinc casting alloys are shown in figure 9.12 and have a wide range of applications in the domestic appliance and automobile industries.

Chapter 10

Molecular structure of polymers

10.1 Introduction

With the increasing use of plastic materials in engineering and their progressive development, it is essential that the student understands and appreciates the molecular structure of polymers as compared with metals.

It may also be of interest to appreciate the fact that plastic materials are not new and that some have been in existance for over a century. Following the initial work carried out by Thomas Hancock, a British inventor, Charles Goodyear, an American, had established a process for vulcanising rubber by the mid-1850s. This was shortly followed by the development and introduction of cellulose plastics. Both of these plastics use natural vegetable products as the basic raw material. The turn of the century saw the introduction of *bakelite*, developed by the Belgian, Dr Leo Baekeland. During the 1920s and 1930s, the vinyl plastics PVC, polyethylene (polyethene) and polystyrene were introduced, with poly-amides—nylons—as a moulding material and PTFE in the early 1940s. By the end of the Second World War the acrylic plastic, Perspex, was also available. Rapid development and introduction of other plastics has continued to the present day in order to meet the demands of industry.

Such interest in polymers results from their excellent electrical and thermal insulation properties and their low specific density. With some polymers, their low frictional resistance and good weight:strength ratio has given impetus to their development, but with all polymers the ease with which they can be moulded and hence the lower cost of the finished product makes polymers one of the most significant developments of the twentieth century.

Figure 10.1 illustrates the stages in the manufacture of some common polymers, from the natural raw material to the finished polymer. It should be noted that other natural raw materials used in polymer manufacture are lime, fluorspar and animal and vegetable products.

10.2 The role of carbon in polymer structures

The structure of all polymers is one of long molecular chains, hundreds of molecules in length. In most cases the molecular chains are entangled, not unlike a wad of cotton wool, and held together as a solid mass by the forces of attraction which exist between molecules in adjacent chains (figure 10.2).

Since polymers are basically carbon compounds of a relatively simple structure, a brief study of their chemistry may provide a better understanding of their various structures in polymers.

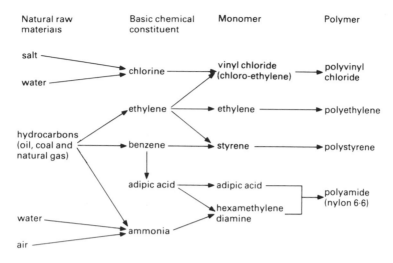

Figure 10.1 Simplified flow chart for polymer production.

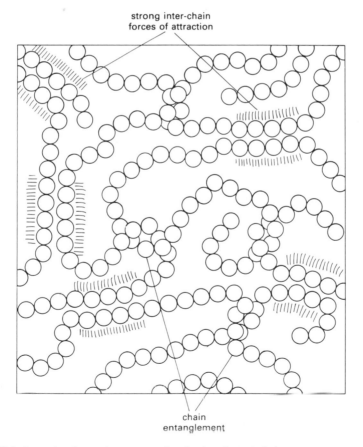

Figure 10.2 Attraction forces between molecules in adjacent chains.

Different elements have different chemical combining capacities, known as valencies or *bonds*—the valencies of hydrogen and chlorine being one, of oxygen two and of carbon four. These valencies can be represented in a chemical formula by dots or dashes. The valencies of atoms can be regarded as the function of the outer shell of electrons surrounding every kind of element. The solution to the problem of valencies of non-metallic elements is found by studying the inert gas argon. Argon has no valency but is surrounded by eight orbiting electrons. This produces a system which has extreme stability. However, a characteristic of other non-metallic elements which have valencies is that they have outer shells with less than eight electrons. They form compounds with one another by 'sharing electrons' in their orbits, each atom in the compound being surrounded by an 'octet' of electrons. It thus has the same structure and stability as argon. An example is carbon which has only four electrons in its outer shell. With such an element there are many ways of forming the octet necessary for stability. An appreciation of this situation will help in understanding the complexity of carbon chemistry and hence polymer structures.

If compounds of carbon and hydrogen are considered, which are elements predominent in polymers, it can be shown in terms of 'shared electrons' that carbon requires four atoms of hydrogen to form a stable molecule.

$$\cdot \overset{\cdot}{\underset{\cdot}{C}} \cdot \; + 4(\dot{H}) \qquad gives \qquad H : \overset{H}{\underset{H}{C}} : H \qquad or \qquad CH_4 \text{ (methane)}$$

Hence 'paired' electrons represent one valency.

The form of bonding taking place through the sharing of electrons is one which is predominent in polymers and is known as a covalent bond and is significant in explaining the ease with which carbon chains of this type are formed.

In the case of the most common hydrocarbon chain molecule in polymers, that derived from ethylene (ethene), the molecule comprises one atom of carbon and two atoms of hydrogen $(CH_2)_{2n}$. In terms of shared electrons, it can be seen that such molecules readily combine to form molecular chains and gives rise to what is known as a *double bond*, i.e. since *paired electrons* represent one valency.

$$\underset{H}{\overset{H}{:C:}} + \underset{H}{\overset{H}{:C:}} \qquad or \qquad \underset{H}{\overset{H}{C:}} \rightleftharpoons \underset{H}{\overset{H}{:C}} \qquad or \qquad \underset{H}{\overset{H}{\diagdown}} C = C \underset{H}{\overset{H}{\diagup}}$$

unstable   achieving stability   double bond

Where a double bond exists, there are two pairs of electrons between the carbon atom nuclei, one having approximately two-thirds the strength of the other. In such cases it is possible to break the double bond by chemical processing. This restores the carbon–carbon structure to the normal single shared electron structure—the single bond—and allows the molecules to join together in a 'chain' structure. An example of 'opening' the double bond is shown as follows: that is, converting C_2H_4 into an active molecule CH_2 with spare valencies capable of joining up with similar 'opened' molecules to form a long chain.

$$\left(\begin{array}{c} H \\ \diagdown \\ C=C \\ \diagup \\ H \quad H \end{array}\right)_n \longrightarrow \left(\begin{array}{c} H \\ | \\ -C- \\ | \\ H \end{array}\right)_{2n} \longrightarrow \sim \begin{array}{ccccc} H & H & H & H & H \\ | & | & | & | & | \\ C-C-C-C-C \\ | & | & | & | & | \\ H & H & H & H & H \end{array} \sim$$

In terms of the octet theory and covalent bond, ethylene can be represented thus:

$$\sim \begin{array}{cccccc} H & H & H & H & H & H \\ \cdot\cdot & \cdot\cdot & \cdot\cdot & \cdot\cdot & \cdot\cdot & \cdot\cdot \\ :C: & C: & C: & C: & C: & C: \\ \cdot\cdot & \cdot\cdot & \cdot\cdot & \cdot\cdot & \cdot\cdot & \cdot\cdot \\ H & H & H & H & H & H \end{array} \sim$$

Hence a very strongly bonded molecular chain can be formed with a carbon backbone.

It follows that long molecular chains can be formed with a very strongly bonded carbon backbone in which hydrogen atoms may be replaced by other single-valent atoms or molecules. It is, therefore, no surprise to learn that most polymers use a hydrocarbon raw material in their manufacture (figure 10.1).

Although the chemical bond between the carbon atoms is permanent, the strength of the forces of attraction between molecular chains can be reduced by the application of heat. This, in effect, increases the vibrations of the molecules within the chains and in consequence increases the distance between them. Hence, any application of force whilst in this condition will more readily effect a change in shape whereby the chains slide over each other with greater ease. It therefore follows that when the heat source is removed, the previously weakened forces of attraction will progressively strengthen and the polymer mass will again become solid and more rigid.

10.3   Terminology in polymer technology

The unstable single repeating unit $-CH_2-$ (which cannot exist by itself as a molecule) as referred to in Section 10.2 is often known as the *mer*. This results from the opening of the double bond in the C_2H_4 molecule and, as a single unit, is known as the *monomer*. The stability of the system is restored during the chemical process known as *polymerisation*—the external agent—in which the monomers are joined together to produce a substance comprising long molecular chains the *polymer*.

The process so described is, in fact, *addition polymerisation*, a process in which the reaction takes place without any by-product. However, there is another type of polymerising process—*condensation polymerisation*—in which a by-product is rejected or condensed out during the reaction. The type of polymerising reaction that takes place depends upon the particular polymer being produced.

Polymeric materials comprising identical molecules in the chain—one type only—are described as *homopolymers*, whereas those which contain chains with two or more different types of molecules are described as *copolymers*.

$$
\text{homopolymers} \quad \sim
\begin{array}{cccccccc}
H & H & H & H & H & H & H & H \\
| & | & | & | & | & | & | & | \\
C-&C-&C-&C-&C-&C-&C-&C \\
| & | & | & | & | & | & | & | \\
X & X & X & X & X & X & X & X
\end{array}
\quad \sim
$$

$$
\text{copolymers} \quad \sim
\begin{array}{cccccccc}
H & H & H & H & H & H & H & H \\
| & | & | & | & | & | & | & | \\
C-&C-&C-&C-&C-&C-&C-&C \\
| & | & | & | & | & | & | & | \\
X & Y & X & Y & X & Y & X & Y
\end{array}
\quad \sim
$$

A further study of polymers will reveal fundamental differences in the overall structure in terms of the type of linkage both along and between the molecular chains. Such differences give rise to the different types of polymers which can be classified as *thermoplastics*, *thermosetting plastics* and *elastomers*.

Thermoplastics (thermo-softening) have a structure which basically consists of monomer joined together through the carbon atom to form long molecular chains with forces of attraction between molecules in adjacent chains. This form of polymer can be repeatedly softened and is conventionally represented thus (M represents the monomers):

$$
\begin{array}{l}
\sim M-M-M-M-M-M-M-M-M \sim \\
\sim M-M-M-M-M-M-M-M-M \sim \\
\sim M-M-M-M-M-M-M-M-M \sim \\
\sim M-M-M-M-M-M-M-M-M \sim \\
\sim M-M-M-M-M-M-M-M-M \sim
\end{array}
$$

Thermosetting plastics (thermo-hardening) have a structure similar to the thermo-softening type but with additional three-dimensional cross-links between adjacent chains, thus anchoring the chains together. It should be noted that the degree of cross-linking varies from one polymer to another. This form of polymer, which cannot be softened, is conventionally represented thus:

$$
\begin{array}{l}
\sim M-M-M-M-M-M-M-M-M \sim \\
\sim M-M-M-M-M-M-M-M-M \sim \\
\sim M-M-M-M-M-M-M-M-M \sim \\
\sim M-M-M-M-M-M-M-M-M \sim \\
\sim M-M-M-M-M-M-M-M-M \sim
\end{array}
$$

Elastomers are a particular group of polymers which consist of very long molecular chains which are folded, coiled and entangled, as well as being subject to a small degree of cross-linking. This irregular arrangement allows considerable reversible extension—elasticity—to take place at normal temperature.

10.4 Common molecular chain arrangements

The type of molecular chain arrangement exhibited by a particular polymer dictates the elastic and thermal properties and to some extent classifies the polymer.

At this stage, a brief study of molecular behaviour in polymers when subjected to external forces or to a change in temperature may provide a better understanding of their elastic and thermal properties.

Consider a polymer which is subjected to an external force. The closer the molecular chains, the denser will be the polymer and the greater will be the interchain forces of attraction. Hence, a greater external force will be required to overcome the internal resistance of the structure. When the external force is sufficiently increased to overcome such internal resistance, the molecular chains will slide over each other and effect a change in shape. In a closely-packed dense molecular chain arrangement the internal resistance—the strength—will be relatively high and the polymer will exhibit moderate change in shape before failure—percentage elongation.

When a polymer is subjected to an increase in temperature, the effect of the heat energy input and its conversion into kinetic energy will influence the elastic and flow properties. The heat energy absorbed by the chain molecules will increase their activity which in turn will increase the distance between the chains and hence reduce the forces of attraction between adjacent chains. In consequence, there will be a reduction in the internal resistance of the polymer and hence a smaller external force will be required to effect a change in shape. It follows, therefore, that if sufficient heat energy is absorbed by the molecules, the flow properties will improve and be sufficient to facilitate moulding or forming during component manufacture, i.e. viscous flow.

Basically, there are three types of molecular chain arrangements, namely *linear*, *branched* and *cross-linked* (figure 10.3).

In a linear-chain arrangement the molecules are joined together end-on with no side branches or bulky side appendages attached to the chains and no cross-linking between the chains, as shown in figure 10.3(a). In a polymer structure, this results in the molecular chains being close together, giving a high density and relatively strong interchain forces of attraction. It follows, therefore, that a relatively large external force would be required to overcome such interchain forces of attraction and allow the chains to slide over each other, the stress induced being high and accompanied by a moderate change in shape before failure. If the same polymer was subjected to a rise in temperature, the heat-energy input would increase the molecular activity, thus increasing the interchain spacing and so reducing the interchain forces of attraction. In consequence, a smaller force would be required to induce plastic flow with a greatly increased

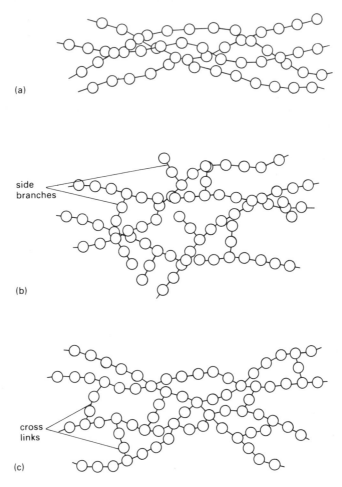

(a)

side
branches

(b)

cross
links

(c)

Figure 10.3 Diagrammatic arrangements of polymer chains. (a) Linear chain arrangement. (b) Branched chain arrangement. (c) Cross-linked chain arrangement.

change in shape before failure. With a further increase in temperature, sufficient viscous flow could be achieved to facilitate moulding.

A branched-chain arrangement is similar to that of linear chains in that branches have developed from the linear chains, as shown in figure 10.3(b). In consequence, such molecular chains in a polymer will be relatively widely spaced giving a low density with relatively weak interchain forces of attraction. Although it may be assumed that a relatively low internal force would be required to induce plastic flow with a significant change in shape before failure, the side branches do have their influence. Such branches increase chain entanglement and restrict the relative chain movement, so increasing the external force necessary to induce plastic flow coupled with a reduction in the change in shape before failure. As with linear chain arrangements, a sufficient increase in temperature would allow viscous flow to take place. As the extent of chain branching increases, there is a noticable decrease in the viscosity, density, melt temperature and yield point. Linear-chain arrangements with bulky appendages exhibit similar characteristics to a branched arrangement. Although there is less chain entanglement, strong forces of attraction usually exist between the bulky atoms or molecules in adjacent chains giving rise to greater strength.

A cross-linked-chain arrangement, as shown in figure 10.3(c), is usually associated with thermosetting plastics whereas linear and branched arrangements are normally found in thermoplastic materials. In the formation of a cross-linked arrangement, during the chemical reaction of condensation polymerisation, the mers join up initially to form a branched-chain structure. In due course these small branched molecules will join up to form a three-dimensional cross-linked arrangement. In practice, it is more convenient to produce a structure of small branched molecules, which are relatively stable at room temperature, as the raw material for subsequent compression moulding operations. These small branched molecules are then deformed during pressure moulding, either under the influence of heat or a catalyst, when they join together and some cross-linking occurs. It should be noted, however, that the extent of cross-linking can vary. As the number of cross-links in a given volume increases, the interchain spacing decreases with a consequent increase in density and a decrease in flexibility and toughness. With maximum cross-linking the polymer becomes inflexible, hard and brittle. Since the cross-links are chemical bonds, any rise in temperature will not break them and hence no plastic or viscous flow can be induced.

Elastomers are a particular type of cross-linked molecular chain arrangement in which the cross-linking is achieved via a different route to that described in the previous paragraph. In these polymers the structure commences with the formation of very long linear chains which are irregularly coiled, folded and entangled, followed by cross-linking through some reactive group. In the case of natural rubber, controlled cross-linking or 'vulcanising' is effected with sulphur. Figure 10.4 shows light cross-linking, i.e. in only a few of the available positions.

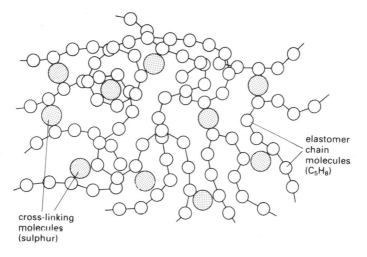

elastomer
chain
molecules
(C_5H_8)

cross-linking
molecules
(sulphur)

Figure 10.4 Diagrammatic arrangement of elastomer chains showing the cross-linking molecules.

When a tensile force sufficient to stretch the material is applied, the molecules partially disentangle and straighten out so that they become orientated in the general direction of the applied force. Since the chain molecules become more closely aligned, the forces of attraction between the molecules in adjacent chains increase so that the elastomeric material becomes stronger and more rigid. However, when the applied force is removed, the molecular chains return to their original shapes. Since cross-links are present, no permanent plastic deformation can occur and only elastic deformation is possible. Figure 10.5 illustrates vulcanised rubber in both an unstressed and a stressed condition.

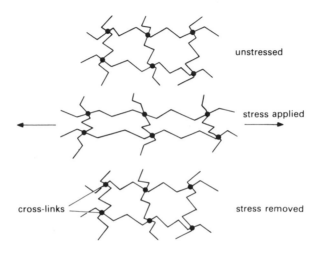

Figure 10.5 Vulcanised rubber before, during and after stressing.

A rise in temperature will increase the molecular activity and in elastomeric materials will result in an increase in the elastic properties and permit some permanent distortion. If there is a decrease in temperature the molecular activity will reduce and so the material will become more rigid and less elastic. This situation may be demonstrated by freezing a golf ball and then attempting to bounce it.

Chapter 11

Effect of structure and additives on the properties of polymers

11.1 Types of structures in thermoplastic polymers

Before consideration is given to the influence of the structural arrangement of polymers on their properties, a brief study of the formation of the giant molecules might be a useful exercise.

The 'solidification' or polymerisation of a polymeric material is not unlike the dendritic solidification of metals described in Chapter 1. In both cases solidification commences from randomly oriented nuclei or growth points within the liquid. As the temperature falls, heat energy is given up with a consequent reduction in the kinetic energy—relative motion—of the atoms. When the energy level within the atoms falls below the value of the *bonding* force offered by *stationary* atoms or molecules, they will join up and so progressively increase the size of the solid mass. This mechanism continues until all the atoms have given up their kinetic energy via heat energy and join up with each other. If the rate of solidification is slow enough—under equilibrium conditions—the structure will assume its natural formation. If, however, the rate of cooling is greater than that required for equilibrium, an 'unnatural' structure will form.

In the case of polymeric materials, the growth from each nuclei is basically linear, the molecular chains forming during the polymerisation process as illustrated in figure 11.1.

Initially the C_2H_4 stable ethylene monomers are charged into the reactor and, as such, are very-short-chain molecules, two carbon atoms long, as shown in figure 11.1(a). At this stage they possess a relatively high degree of mobility and hence the interchain forces of attraction, the *van der Waal's forces*, are weak and ineffective. As the polymerisation reaction proceeds, the double C–C bond is opened and the active unstable mers join up with other similar mers with a covalent C–C bond (see Section 10.2). Thus the molecules grow in length as shown in figure 11.1(b). In consequence, the van der Waal's forces increase in their effectiveness, thereby reducing the activity of individual molecules and rendering the material relatively viscous. With the completion of the polymerisation reaction, the chain molecules will have increased their length to such an extent that the magnitude of the van der Waal's forces acting between adjacent molecular chains renders the material so viscous that it may be considered as a solid. Such a condition is illustrated in figure 11.1(c) in which polyethylene molecules consist 'on average' of a chain with approximately 1200 carbon atoms in length.

161

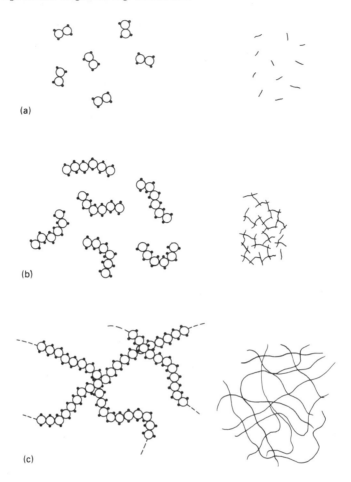

Figure 11.1 Diagrammatic representation of the stages in the polymerisation of ethylene (a) Two carbon atom chains. (b) After some growth. (c) Van der Vaal's forces where chains are approximately 1200 atoms long.

It should be noted that the number of carbon atoms in the chains dictates the molecular weight and since the chain lengths vary, so does the molecular weight. In polymer technology, it is accepted practice to refer to average molecular weights rather than the molecular chain lengths. Since polymers of a high *average* molecular weight, i.e. relatively long chains, are the most useful to the engineer, it is for this reason that those used in industry are referred to as *high polymers.*

Within a polymeric material, the molecular chain structure may assume different arrangements, depending upon the particular chain pattern and whether equilibrium conditions prevailed during their formation.

If the natural molecular chain formation of a polymer is one which has large side appendages, either branches or bulky molecules, there is likely to be a very high degree of chain entanglement. In such cases the molecular chains can be said to be in random disordered array, a structure generally described as *amorphous* (figure 11.2).

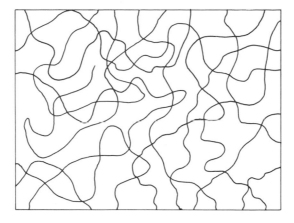

Figure 11.2 Diagrammatic representation of an amorphous structure.

In some polymers, the natural chain formation is linear, with no branches or bulky side appendages. When this situation arises, it is likely that the chains display considerably less entanglement and so exhibit large regions in which the chain molecules are closely folded and aligned in orderly array. Such ordered regions are known as *crystallites*, the overall structure being described as *crystalline* (figure 11.3). Such regions are not unlike the structure of metals in which the crystals or grains consist of atoms arranged in an orderly regular geometric pattern as described in Chapter 1. However, in polymeric materials, and unlike metals, the degree of crystallinity is not as high and both amorphous and crystalline regions exist side by side, the crystallites being surrounded by amorphous regions. Such materials are known as *semi-crystalline* polymers. Figure 11.4 shows the isolated crystallites in an amorphous matrix, in which the extent of any one crystallite will be in the order of 0.025 μm.

Figure 11.3 Diagrammatic representation of a crystalline structure.
Note: It is highly unlikely that a 100% crystalline would form in polymer.

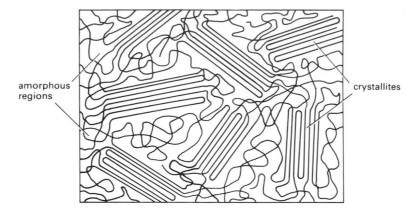

Figure 11.4 Diagrammatic arrangement of a semi-crystalline structure.

In addition to the particular type of polymerising process used, the degree of crystallinity can be influenced by the rate of cooling following polymerisation or processing. If the polymer is cooled under non-equilibrium conditions, the development of large crystalline regions is restricted. This situation can be compared with the quench hardening of steels as described in Chapter 5. In some cases the degree of crystallinity may be as high as 90%, but is generally in the order of 50%.

11.2 The effect of structure on mechanical properties

Branched and bulky side-appendaged molecular chains, by their very nature, produce a highly amorphous structure. Since the branches or bulky side molecules prevent the molecular chains from becoming closely packed, there is inevitably a high degree of entanglement.

Where a branched structure is evident the van der Waal's forces will be relatively weak and so permit plastic flow to take place. However, the side branches limit the extent of plastic flow due to entanglement and hence will increase the internal resistance of the structure. It follows, therefore, that the strength of the material will be relatively low with moderate elasticity. The greater spacing between adjacent molecular chains will also influence the toughness, such structures having a relatively high value and so display a measure of flexibility. In the case of a linear structure with bulky side atoms or molecules, the structure is similar to that of branched chains. However, there are additional very strong forces of attraction in existance between the bulky side appendages which tend to increase the density and strength but reduce the toughness and flexibility.

When considering a linear molecular chain structure, there is much less entanglement and more likelyhood of the chain molecules being arranged in a closely packed, folded arrangement giving a high density polymer. In such cases the van der Waal's forces will be very high which will be reflected in the high strength of the material. Although the elasticity will be relatively high, the polymer will exhibit little flexibility and toughness. Such are typical properties of a highly crystalline polymer.

When both amorphous and crystalline structures develop together, as in semi-crystalline polymers, their strength must be influenced by the degree of crystallinity. Since crystalline regions display high strength with little toughness and amorphous regions display low strength but a high degree of toughness, a semi-crystalline polymer will be relatively strong with a reasonable measure of toughness. However, as the degree of crystallinity increases, the stronger will be the polymer, at the expense of toughness.

11.3   Additives in polymers, their purpose and applications

Any additive compounded with a basic polymer will influence the structure and hence the properties. Such additives may be used to stabilise the polymer, to improve the working properties, to change the mechanical or physical properties or to introduce colour. Such additives may be identified as stabilisers, plasticisers, fillers or colourants. Although colourants do not affect mechanical properties, the authors consider their inclusion in this section is essential in order to give a full picture of the range of additives used with polymers.

Stabilisers

Some polymers deteriorate when exposed to heat or light or during moulding operations, whereas some weather in the final moulded form. This deterioration may be the result of a partial breakdown of the polymer itself or from residual or atmospheric oxidation. Examples of heat stabilisers range from lead and tin compounds to barium, cadmium and zinc soaps. Carbon black is used as an ultra-violet stabiliser, particularly in vinyl polymers.

Polyvinyl chloride is a particular polymer which suffers deterioration and hence requires stabilising. With untreated rigid vinyl chloride polymers a structural change takes place at temperatures above 70°C, together with a colour change. The structural change results from hydrogen chloride gas being liberated which leads to a degradation of the polymer, i.e.

With dehydrochlorination, polyene structures are formed but the visible evidence of such degradation is one of a colour change. Initially water-white, on heating it will turn pale yellow at around 70°C, through orange and brown to black. Further degradation causes adverse changes in mechanical and electrical properties. Thus in practice, changes in colour on heating provide a simple and readily obtainable criterion of degradation. When considering the working of plasticised PVC, it must be heated to a temperature between 150°C and 200°C. At this temperature further degradation and discolouring of the material takes place as a result of further gas being liberated, giving rise to a hard, brittle material.

When PVC is exposed to bright sunlight polymer decomposes, in this case as a result of oxidation. The effects are similar to that from heating but to a lesser degree.

Hence, it follows that stabilisers for PVC should possess the following characteristics: (a) to combine with and 'fix' the gases which are liberated with a rise in temperature; (b) to prevent decomposition as a result of oxidation; and (c) to filter the ultraviolet rays from sunlight.

Plasticisers

These additives are either high-boiling liquids, usually esters or low-melting solids such as camphor, which are added to polymers to modify their physical properties. Originally, as the name implies, their primary purpose was to render the polymer more plastic. However, they are frequently necessary to improve flexibility and sometimes to promote better moulding properties such as improved flow during fabrication.

Solvents may also be used in the initial compounding of the polymer. Since the solvents are volatile and vaporise, the properties imparted by them are only temporary. Plasticisers, however, if properly selected, impart permanent properties to the product.

In addition to improving flexibility, plasticisers may modify one or more of the following characteristics: elongation, tensile strength, toughness, softening point, melting point, heat-sealing properties, flow properties, flame resistance, electrical properties, water absorption, oil resistance and toxicity.

In considering the mechanism in which a plasticiser acts, both solvents and plasticisers can penetrate between the molecular chains, thus reducing the van der Waal's forces, as shown in figure 11.5. This modifies the properties of the polymer, the extent of which depends on the magnitude of the original van der Waal's forces holding the chains together.

Plasticisers with the assistance of heat and sometimes solvents are required to plasticise such strongly bonded chains as those in PVC. The polymer–polymer bonds may be replaced with polymer–plasticiser bonds, thus not only producing flexibility but additional toughness. It is possible in some cases to add sufficient plasticiser so that it forms a continuous phase in which the chains move. Such a plasticised mass may be a soft gel.

The following factors need to be considered when selecting a plasticiser.

(1)   Low volatility: if the plasticiser evaporates in a short period of time under use, the product will suffer deterioration in desirable properties.

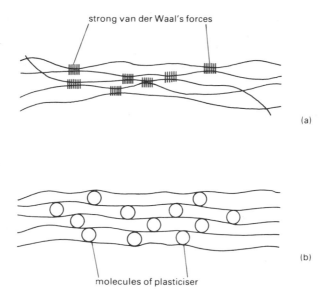

Figure 11.5 Effect of plasticisation on polymer chain molecules. (a) Before and (b) after addition of plasticiser.

(2) Toxicity: some plasticisers produce dermatitis upon contact with the skin. Others may produce internal toxic effects from contact with food or drink.

(3) Solubility: some plasticisers may be soluble in solvents with which the finished product may come into contact, again with deterioration in desirable properties.

(4) Stability: some plasticisers may reduce the stability of the polymer in respect of moisture, heat and light.

(5) Migration: certain plasticisers may migrate into other materials with which they come into contact, particularly other plastics, which may produce adverse effects.

(6) Corrosion: some plasticisers may induce serious corrosion problems if the polymer comes into contact with metals.

Fillers

Fillers are used primarily to improve the properties of the finished product. They frequently reduce brittleness and increase impact strength. In most cases the effect is mechanical, the fillers not reacting with the polymer resin. Usually the fillers are cheaper than the resin and as a result they reduce the overall cost.

Most polyethylenes used in industry do not contain any filler material. However, when fillers are used their function is to improve the physical properties such as flame retardation, anti-static, resistance to weathering and to oxidation. In such cases, the fillers are in the form of oxides, esters, carbon black and amines.

With PVC, fillers are generally only used in opaque compounds. In order to reduce costs, a general-purpose filler of calcium carbonate is used, whereas asbestos is used to improve the hardness in floor coverings, and china clay to improve electrical insulation properties.

Polystyrene is normally rigid and hard with low-impact strength. The impact strength can, however. be improved by incorporating a rubbery material into the polymer, thus increasing the toughness. In this case butadiene is used in quantities ranging from 5% to 20%, at the expense of mouldability, working temperature and surface finish.

In the case of polyamides (nylons) the filler used is usually in the form of a reinforcement, glass fibres being the most commonly used. Such additions can double the ultimate strength. When circumstances demand a reduction in the frictional resistance of a polyamide, molybdenum disulphide may be added.

Colourants

These may be dyes, inorganic or organic pigments. In selecting colourants for a specific polymer, consideration must be given to factors other than the actual colour itself. Some colourants are rendered unstable when the finished product comes into contact with certain chemicals, some are unstable under acid conditions whereas others become unstable under alkaline conditions. In some cases colourants may be sensitive to oxidation or reduction, depending upon the conditions to which the polymer is subjected. Excessive heat during moulding or in use or prolonged exposure to ultraviolet light may also cause a breakdown of some colourants unless a suitable stabiliser is used.

Dyes are organic chemical compounds used mainly for colouring transparent and translucent polymers. They are available in a wide range of colours and have excellent brightness and low plasticiser absorption qualities. However, they have poor resistance to both heat and light and they have a tendency to bleed.

Inorganic pigments are mainly metallic oxides which are much superior to organic colourants in weather resistance, in stability to heat and light and in their resistance to bleeding. They are available in a limited range of colours which unfortunately are weak tinctorially. Their use is restricted to opaque polymers such as phenolics.

Organic pigments, in general, have a limited solubility and high oil and plasticiser absorption qualities. They are available in many colours which have good brightness and high colouring strength. Their resistance to heat and light is better than that of dyes but poorer than that of inorganic pigments. Their application is in practically all polymers except silicons.

Chapter 12

Common polymeric materials

Having studied the molecular and structural variations in polymeric materials and their influence on the properties, it is now possible to consider specific thermoplastic polymers in terms of characteristics, structure and hence properties, from which justification of their applications can be made. The four specific polymers considered in this chapter are those in most common use today. It is of interest to note that these polymers are the highest tonnage produced today. For simplicity, the values of properties quoted are typical, even though a range may apply depending upon variations which can occur due to molecular chain structural variations or the presence of additions.

12.1 Polyethylene

This is the parent of all vinyl polymers (see figure 10.1). The discovery and development of polyethylene provides an excellent example of the value of observing and following up an unexpected experimental result. In 1931, the research laboratories of the Alkali Division of the Imperial Chemical Industries (ICI) designed an apparatus to investigate the effect of pressure up to 3000 bars on binary and ternary systems. During the research programme it was noticed that in one of the experiments in which ethylene gas (C_2H_4), a by-product from petroleum refining, was being used, a small amount of white, waxy solid had formed. On analysis it was found to be *polyethylene*, a homopolymer produced by addition polymerisation.

$$\left(\begin{array}{c} H \quad\quad H \\ \diagdown\,C{=}C\,\diagup \\ \diagup \quad\quad \diagdown \\ H \quad\quad H \end{array}\right)_n \longrightarrow \left(\begin{array}{c} H \\ | \\ {-}C{-} \\ | \\ H \end{array}\right)_{2n} \longrightarrow \ \sim\!\! \begin{array}{c} H\ \ H\ \ H\ \ H\ \ H\ \ H\ \ H\ \ H \\ |\ \ \ |\ \ \ |\ \ \ |\ \ \ |\ \ \ |\ \ \ |\ \ \ | \\ C{-}C{-}C{-}C{-}C{-}C{-}C{-}C \\ |\ \ \ |\ \ \ |\ \ \ |\ \ \ |\ \ \ |\ \ \ |\ \ \ | \\ H\ \ H\ \ H\ \ H\ \ H\ \ H\ \ H\ \ H \end{array} \!\!\sim$$

monomer	mer	polymer
(ethylene)		(polyethylene)

Subsequently, attempts were made to reproduce this polymer, initially without success. However, it was eventually discovered that a trace of oxygen was necessary to bring about the formation of the polymer and its source in the initial experiment was found to be quite accidental—a leak being found in the apparatus.

Polyethylene was first produced at ICI by this high-pressure system on a commercial basis in September 1939 and analysis revealed that the chain molecules produced were branched and the polymer had a relative density of

0.92. Such a structure was found to have a relatively low degree of crystallinity, about 50%.

```
                              H
                              |
                         H—C—H
                              |
                         H—C—H
                              |
                         H—C—H
   H   H   H   H   H   H   H   H   |   H   H   H
   |   |   |   |   |   |   |   |   |   |   |   |
~ C—C—C—C—C—C—C—C—C—C—C—C ~
   |   |   |   |   |   |   |   |   |   |   |   |
   H   H   H   |   H   H   H   H   H   H   H   H
              H—C—H
                |
              H—C—H
                |
              H—C—H
                |
                H
```

In the mid-1950s a polyethylene of a higher density became available, the relative density being of the order of 0.96. When the structure of this polymer was analysed the chain molecules were found to be unbranched, i.e. linear, with a high degree of crystallinity, of the order of 90%. Hence, density is an indication of the degree of crystallinity and so for technological purposes, density can be taken as a measure of the degree of short-chain branching.

Low-density polyethylene

Polymerisation is carried out at high pressures, between 1000 and 3000 bars and at temperatures of between 80°C and 300°C. It is carried out by passing ethylene gas through an oxidising catalyst or initiator such as benzoyl peroxide. The polyethene is extruded as a continuous band, solidified by passing through a water bath and then cut into granules for subsequent forming.

Due to the side branching of the chain molecules, crystallinity is limited to about 50%. Hence the structure comprises isolated crystallites in an amorphous matrix (see figure 11.4). Low-density polyethylene may therefore be described as a semi-crystalline polymer. Since the side branching also increases entanglement and decreases the van der Waal's forces, the polymer will be tough and flexible with low strength and moderate elasticity.

The polymer is white, translucent, waxy, odourless and tasteless. It is tough and flexible over a wide range of temperatures, even down to $-75°C$. Its soft surface is easily marked, it is easily moulded and machined and has good dimensional stability. Low-density polyethylene has excellent resistance to common solvents and chemicals, it has good weathering properties and excellent electrical insulation properties, even in humid conditions. However, the polymer suffers slight deterioration with prolonged exposure to heat and light. Typical properties of low density polyethylene are: relative density, 0.92; ultimate strength, 11 N/mm²;

percentage elongation, 100%; maximum service temperature, 85°C. *Note*: as crystallinity increases, density increases with a corresponding increase in service temperature, rigidity, strength and hardness.

Low-density polyethylene, a low-cost, easily moulded, general-purpose polymer has industrial and domestic applications in packaging (i.e. bags and squeeze-bottles) and electrical insulation as a coating on cables and wires.

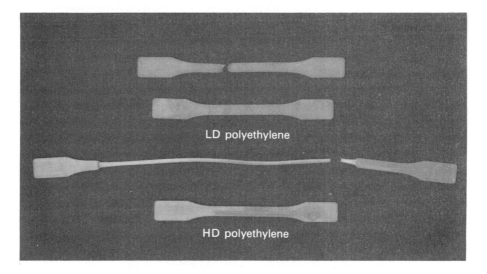

Figure 12.1 Tensile-tested specimens of low-density and high-density polyethylene to illustrate the difference in percentage elongation.

High-density polyethylene

Polymerisation is carried out at low pressure (very little above atmospheric) and at a temperature of 70°C. The essential difference between the high-pressure process and the low-pressure process is the catalyst used. For high-density (low-pressure) polyethylene the catalyst or initiator is a combination of aluminium tri-ethyl and titanium tetrachloride suspended in a light hydrocarbon spirit.

Since the chain molecules are linear with no side branching the structure will display less chain entanglement and stronger van der Waal's forces as compared with low-density polyethene. This will promote a more crystalline structure with the crystallinity rising as high as 90% in some cases. As a result of such a structure the polymer will possess increased rigidity, strength and hardness at the expense of toughness and flexibility.

The characteristics displayed by high-density polyethylene are the same as those for low-density except that it is opaque and it possesses increased resistance to chemicals and solvents. Typical properties are: relative density, 0.96; ultimate strength, 31 N/mm^2; percentage elongation, 350%; maximum service temperature, 125°C.

Industrial applications utilise the improved properties of rigidity, higher resistance to chemicals and higher service temperature in piping, chemical, electronic, and medical equipment, and in a wide range of domestic goods.

12.2 Polyvinyl chloride (PVC)

This copolymer is produced by addition polymerisation of the monomer vinyl chloride, which may be prepared by two different methods. For many years the major route was by the addition of hydrochloric acid to acetylene but with the development of the petrochemical industry in recent years there has been a swing from acetylene to ethylene. Hence the majority of vinyl chloride is produced from ethylene gas and chlorine (sometimes described as chloro-ethylene), the double-bonded monomer then being polymerised.

The most common method of polymerisation is by the suspension technique which is easier to control with little loss of clarity or electrical-insulation properties. The monomer, a gas which liquifies easily under pressure, is suspended in demineralised water in the form of droplets. Such droplets are kept from coalescing by the addition of a suspension agent such as gelatine. A catalyst is added which is soluble in the monomer droplets but insoluble in water and the mixture is stirred at a temperature of 50°C. In the reactor, the temperature induces a rise in pressure, initially to 7 bars, which falls to just above atmospheric as the reaction proceeds. The polymer, in the form of a slurry with water is vented off, cooled and dried to provide polyvinyl chloride powder.

| ethylene + chlorine | monomer (vinyl chloride) | mer | polymer (polyvinyl chloride) |

X-ray analysis of commercially produced PVC indicates that the structure is substantially amorphous, although about 5% crystallinity may exist (figure 11.2). The formation of the amorphous structure is due to the bulky chlorine atoms in the chain molecules. It is of interest to note that the relative atomic mass of chlorine is 35 as compared with 12 for carbon and 1 for hydrogen. A diagrammatic arrangement of the structure of PVC is shown in figure 12.2.

If the bulky chlorine atoms are arranged in a random manner on alternate sides of the linear molecular chains and there are strong forces of attraction between the chlorine atoms, the polymer will possess characteristics similar to that of a polymer with a branched molecular structure. However, polyvinyl chloride resin is difficult to process due to its thermal instability. In order that the polymer can be processed, it must be compounded with suitable additives (see Section 11.3). Such additives improve thermal stability and increase flexibility at the expense of strength.

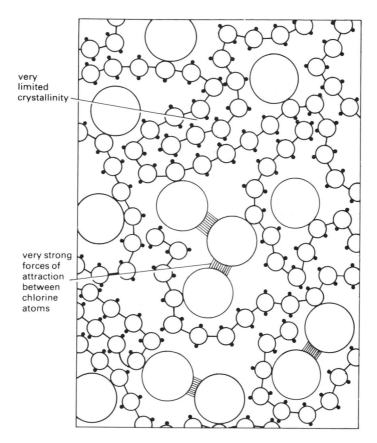

Figure 12.2 Diagrammatic arrangement of the amorphous structure of polyvinyl chloride, illustrating the relative sizes of the chlorine, carbon and hydrogen atoms in the molecular chains.

Polyvinyl chloride is available in both rigid and flexible forms. Unplasticised, rigid PVC is tough and hard with good dimensional stability, minimum water absorption capacity and good resistance to acids, alkalis and most common solvents. However, when a plasticiser is introduced the flexibility and surface finish improves with the polymer becoming softer and rubbery, but at the expense of chemical resistance. Plasticised polyvinyl chloride unfortunately stiffens with age. In both forms the polymer has good electrical insulating properties, particularly in low-frequency ranges.

Typical properties of rigid and plasticised polyvinyl chloride are: relative density, rigid 1.37, plasticised 1.35; ultimate strength, rigid 50 N/mm^2, plasticised 15 N/mm^2; percentage elongation, rigid 15%, plasticised 300%; maximum service temperature, rigid 70°C, plasticised 80°C. *Note*: the properties of plasticised PVC will vary considerably depending upon the degree of plasticisation and the additives used.

Rigid PVC has a wide range of applications depending upon the requirements. For ducting and chemical tank linings where corrosion resistance is required. For water drainage piping, corrugated roof sheeting and safety helmets where rigidity is required. For piping in the gas supply industry where non-porosity is required.

Flexible PVC is used in applications which utilise the improved flexibility and surface finish such as artificial leather cloth, protective clothing, packaging and cable-insulating covering.

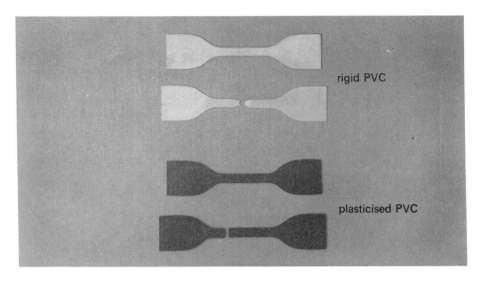

Figure 12.3 Tensile-tested specimens of rigid and plasticised polyvinyl chloride to illustrate the difference in percentage elongation .

12.3   Polystyrene

Although the true nature of the polymerisation of polystyrene was recognised in 1845, some 80 years elapsed before commercial production of the styrene monomer began simultaneously in Germany and the USA in 1925. By 1939 both styrene and polystyrene were well-established products although production rates were low. The greatest demand came during the Second World War when the Japanese cut off the supply of natural rubber and large quantities were required for the production of styrene/butadiene rubbers. With the renewed availability of natural rubber at the end of hostilities, demand for synthetic rubber dropped with a surplus capacity of plant together with an abundance of knowledge and experience gained. This resulted in the production of a cheap, general-purpose polymer which is now one of the major thermoplastic materials.

Styrene, sometimes described as vinyl benzene, is an aromatic vinyl compound which readily copolymerises by addition, even without a catalyst or heat although the process may be accelerated by the use of a catalyst such as benzoyl peroxide and the application of heat. The styrene monomer is made from phenyl (C_6H_5), derived from benzene together with ethylene, both of which are obtained from either oil or coal. The monomer produced is a colourless liquid which polymerises into a transparent glass-like plastic.

$$\left(\begin{array}{c}H \\ \diagdown \\ C=C \\ \diagup \\ H \end{array}\begin{array}{c} H \\ \diagdown \\ H \end{array}\right) \xrightarrow{C_6H_5} \left(\begin{array}{c}H \\ \diagdown \\ C=C \\ \diagup \\ H \end{array}\begin{array}{c} H \\ \diagdown \\ C_6H_5 \end{array}\right) \longrightarrow \left(\begin{array}{c} H \quad H \\ -C-C- \\ H\ C_6H_5 \end{array}\right) \longrightarrow \sim \begin{array}{c} H \quad H \quad H \quad H\ C_6H_5\ C_6H_5\ H \\ C-C-C-C-C-\quad\quad C-C \\ H\ C_6H_5 H \quad H \quad H \quad\quad H \quad H \end{array} \sim$$

ethylene + phenyl   monomer   mer   polymer
               (styrene)   (polystyrene)

Polymerisation is commonly carried out by *mass, solution* or *suspension* techniques. Mass polymerisation has the advantage of simplicity, giving a polymer of high clarity and very good electrical insulation properties. Following polymerisation the melt is extruded as filaments which are cooled, disintegrated and packed.

Examination of the polymer reveals that the structure is substantially linear with bulky side molecules. It should be noted that the phenyl molecule is approximately 40 times the mass of a hydrogen atom. However, because of the random spatial position of the benzene ring, it is incapable of crystallisation, hence, the polymer is entirely amorphous (figure 12.4).

As with PVC the bulky side appendages stiffen the polymer chains and also induce relatively strong forces of attraction between the phenyl molecules in adjacent chains. In consequence, the resin is hard at room temperature. As a result of the relatively high intermolecular forces between the benzene rings the polymer will possess high-tensile properties together with the associated brittleness, although these will vary with the variations in the methods of polymerisation. Polystyrene resin possesses good mouldability, dimensional stability and electrical insulation properties which it retains at low temperatures. However, its limitations are brittleness, low service temperature and mediocre oil resistance. Typical properties of polystyrene resin are: relative density 1.05; ultimate strength 40 N/mm^2; percentage elongation, 2%; maximum service temperature 75°C.

Polystyrene resin is used in applications which utilise the fact that the properties are retained at low temperatures and for its good optical properties, which give it a particularly high brilliance. Such applications are refrigeration equipment (i.e. door liners and trays) and for display figures and toys.

Since this low-cost resin is too brittle for many applications, manufacturers have made a number of attempts to modify the resin in an effort to improve the toughness. The most successful method was found to be copolymerisation with either butadiene or acrylonitrile or with both.

The *styrene-butadiene copolymer* (rubber-modified styrenes) show a significant improvement in toughness and it must be appreciated that since proportions of the constituents can differ, so can the resulting properties. These *styrene-butadiene rubbers* (SBR) with a 30% styrene content are the world's most-used synthetic rubber.

A further development has been in the blending of SBR with the styrene monomer and then polymerising the styrene in the usual way, thus effecting a significant increase in the impact strength at the expense of tensile strength and maximum service temperature. As can be expected, variations in the properties of these SBR-polystyrene blends provide a range of polymeric materials.

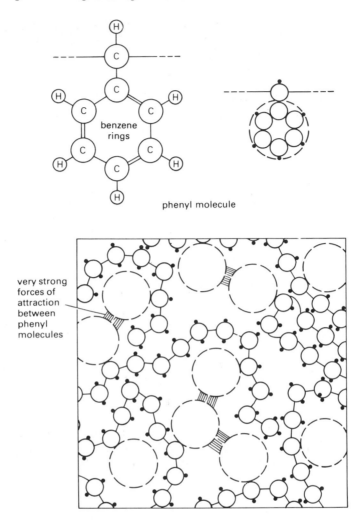

benzene
rings

phenyl molecule

very strong
forces of
attraction
between
phenyl
molecules

Figure 12.4 Diagrammatic arrangement of the amorphous structure of polystyrene, illustrating the relative sizes of the carbon and hydrogen atoms and the phenyl molecules in the chains.

Styrene-acrylonitrile copolymers, with a 25% acrylonitrile content have been available for some time but are rather expensive. Their advantages are improved toughness, service temperature and chemical resistance but unfortunately present great difficulty in moulding and there is a tendency for the resin to discolour towards yellow.

A more recent development in styrene polymers is in the use of both acrylonitrile and butadiene with styrene to form a very tough polymeric material known as an *ABS copolymer*. A typical composition would consist of two parts of a copolymer of 70% styrene and 30% acrylonitrile to one part of a rubber of 65% butadiene and 35% acrylonitrile. In such an ABS copolymer the contents would be approximately 46% styrene, 22% butadiene and 32% acrylonitrile.

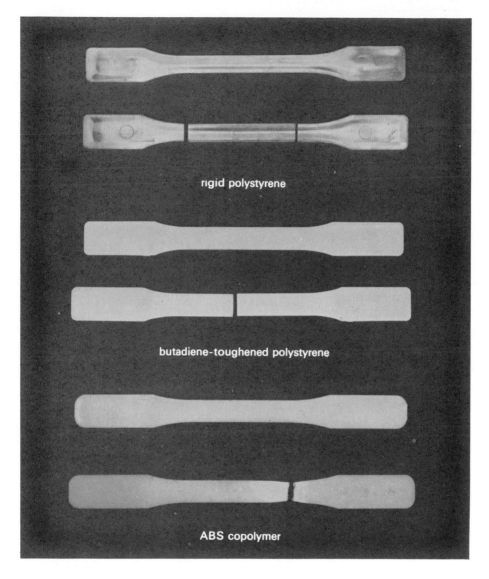

rigid polystyrene

butadiene-toughened polystyrene

ABS copolymer

Figure 12.5 Tensile-tested specimens of styrene-based polymers to illustrate the difference in percentage elongation.

A comparison of the properties between a general-purpose polystyrene resin and the various foregoing modified copolymers is shown in figure 12.6.

12.4 Polyamides (nylons)

Early development of the Nylon 66 was carried out by the Du Pont Company in America in the mid-1930s when Nylon 66 was first synthesised. Commercial

	General purpose polystyrene	Medium impact SBR polystyrene blend	High-impact SBR polystyrene blend	AS copolymer	ABS copolymer
Specific gravity	1.05	1.05	1.05	1.06	1.05
Impact strength (J)	0.25	1.0	2.0	1.0	12.0
Tensile strength (N/mm²)	30	35	20	70	35
Max. working temperature (°C)	65	85	80	100	100

Figure 12.6 Table of properties of polymeric materials based on styrene

production of this polymer for subsequent conversion into fibres was commenced in 1939 but the first nylon mouldings were not produced until 1941. However, the polymer did not become widely used as a moulding material until 1950. Also during the 1930s German chemists investigated a wide range of fibre-forming polymers which resulted in the introduction of Nylon 6. Subsequent research and development has taken place on both sides of the Atlantic, including Russia, resulting in a wide range of condensation polymerised materials being available. It is of interest to note that the first two nylons to be produced still account for nearly all that produced for fibre production.

Nylon is a synthetic thermoplastic polymer described as a polyamide. The term *polyamide* denotes that the polymer consists of molecular chains in which the carbon backbone is interspaced with amide groups of molecules. These semi-crystalline polymers differ from other thermoplastics in that they have atoms other than carbon in the molecule backbone. The copolymer, which is produced by condensation polymerisation (see Section 10.3) has various chain arrangements, depending on the constituents from which it is made.

Nylon polyamides of different characteristics may be made by polymerising a diamine and a dibasic acid of different chemical structures, i.e. hexamethylene diamine and sebacic acid. A simple nomenclature has been adopted to differentiate between the different nylons now being produced. The number of carbon atoms in the constituents of the nylon are indicated by the appropriate figures, the diamine being considered first. In the case of nylons being produced by the polymerisation of a single constituent, the nylon will have a single number: i.e. Nylon 6, polymerised from caprolactam; Nylon 6.6, polymerised from hexamethylene diamine and adipic acid; Nylon 6.10, polymerised from hexamethylene diamine and sebasic acid; Nylon 11, polymerised from amino-undecanoic acid. These are the main molecular combinations of commercial importance although others such as Nylon 7, Nylon 8, Nylon 9, Nylon 10, Nylon 12 and combinations to form Nylon 66/610/6 and Nylon 66/610 are available.

All the chain structures consist of groups of CH_2 molecules interspaced with amide groups (HN–CO) which produces short carbon chains within the linear molecule. The molecular chain structure can be represented in the general form amide group   amide group

$$\sim\left(\begin{array}{c}H\\|\\C\\|\\H\end{array}\right)_x\!\!-\!\!\underset{\underset{O}{\|}}{C}\!-\!N\!-\!\left(\begin{array}{c}H\\|\\C\\|\\H\end{array}\right)_y\!\!-\!\!\underset{\underset{H}{|}}{\overset{\overset{O}{\|}}{C}}\!-\!N \sim + \text{ a condensate}$$

where x and y vary depending on the constituents from which the polymer is made.

Nylon 6

The amide group of molecules occurs after every fifth molecule of CH_2 in the chain. Hence the chain molecules are of the form $[NH(CH_2)_5CO]_x$, which is a repeating unit. When expanded the chain molecules can be expressed as:

$$\sim\left(\begin{array}{c}H\\|\\C\\|\\H\end{array}\right)_5\!\!-\!\!\underset{\underset{O}{\|}}{C}\!-\!N\!-\!\left(\begin{array}{c}H\\|\\C\\|\\H\end{array}\right)_5\!\!-\!\!\underset{\underset{H}{|}}{\overset{\overset{O}{\|}}{C}}\!-\!N\!-\!\left(\begin{array}{c}H\\|\\C\\|\\H\end{array}\right)_5 \sim$$

Figure 12.7 illustrates the intermolecular bonding of Nylon 6.

Although van der Waal's forces exist between chain molecules, the strongest interchain bond is between the amide group of molecules in adjacent chains. This is through the hydrogen atom in the amino molecule (NH) and the oxygen atom in the CO molecule and is highlighted in figure 12.8. Such bonding is often referred to as *hydrogen bonding*.

Nylon 6.6

The substitution of an amide group occurs after six molecules of CH_2 and then after four molecules of CH_2 in the chain, but the arrangement of the amide groups alternate in the positions of the NH and CO molecules. The polymer can be expressed as $[NH(CH_2)_6NHCO(CH_2)_4CO]_x$, which again is a repeating unit. When expanded, the chain molecules can be expressed as:

$$\left(\begin{array}{c}H\\|\\C\\|\\H\end{array}\right)_4\!\!-\!\!\underset{\underset{O}{\|}}{C}\!-\!N\!-\!\left(\begin{array}{c}H\\|\\C\\|\\H\end{array}\right)_6\!\!-\!N\!-\!\overset{\overset{O}{\|}}{C}\!-\!\left(\begin{array}{c}H\\|\\C\\|\\H\end{array}\right)_4\!\!-\!\!\underset{\underset{O}{\|}}{C}\!-\!N\!-\!\left(\begin{array}{c}H\\|\\C\\|\\H\end{array}\right)_6$$

Figure 12.9 illustrates the intermolecular bonding in Nylon 6.6.

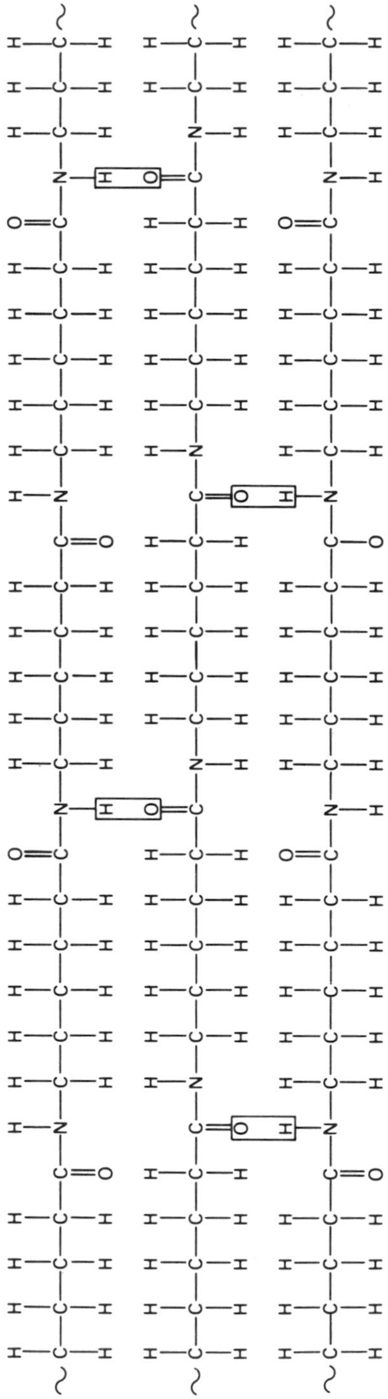

Figure 12.7 Intermolecular bonding of Nylon 6.

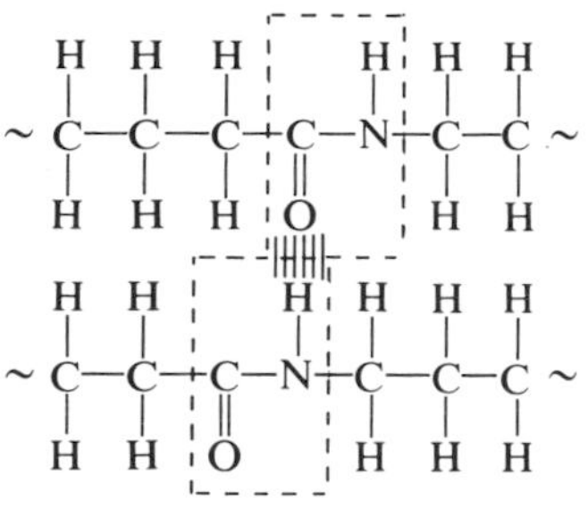

Figure 12.8 Hydrogen bonding.

Nylon 6.10

The substitution of an amide group of molecules occurs after six molecules of CH_2 and then after eight molecules of CH_2 in the chain. The amide group alternation is the same as for Nylon 6.6 and the polymer can be expressed as $[NH(CH_2)_6NHCO(CH_2)_8CO]_x$, which again is a repeating unit. When expanded, the chain molecules can be expressed as:

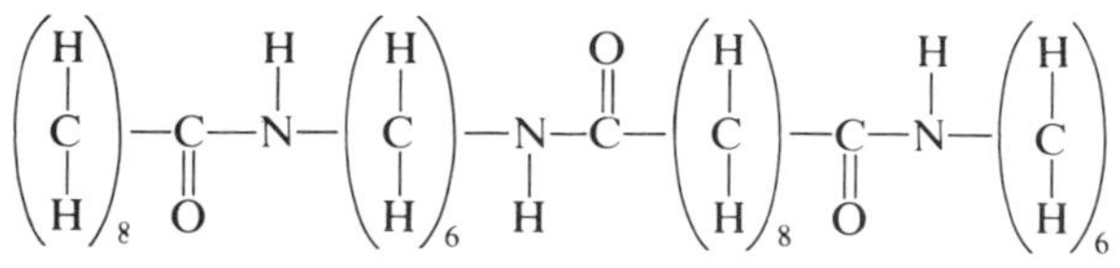

Figure 12.10 illustrates the intermolecular bonding in Nylon 6.10.

Nylon 11

The substitution of the amide group of molecules occurs after every tenth molecule of CH_2 in the chain. Hence, the polymer can be expressed as $[NH(CH_2)_{10}CO]_x$, which again is a repeating unit. When expanded, the chain molecule can be expressed as:

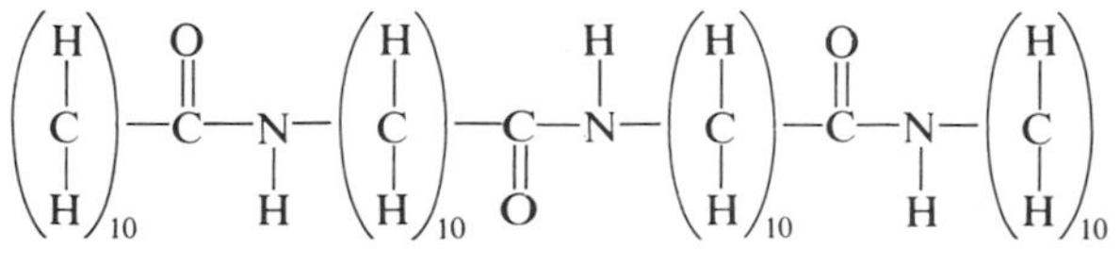

Figure 12.11 illustrates the intermolecular bonding in Nylon 11.

The difference in the resulting properties exhibited by these four types of polyamides stems from the difference in structure of the chain molecules and the difference in bonding between adjacent chains. From figures 12.7 and 12.9 to

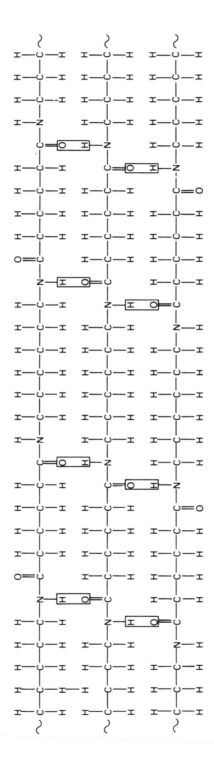

Figure 12.9 Intermolecular bonding of Nylon 6.6.

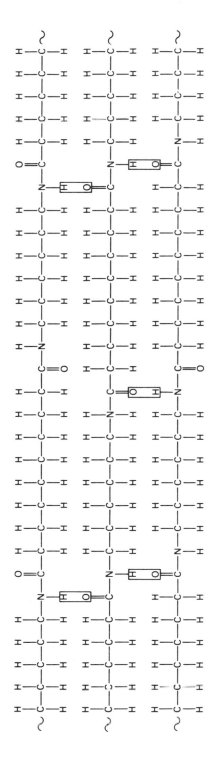

Figure 12.10 Intermolecular bonding of Nylon 6.10.

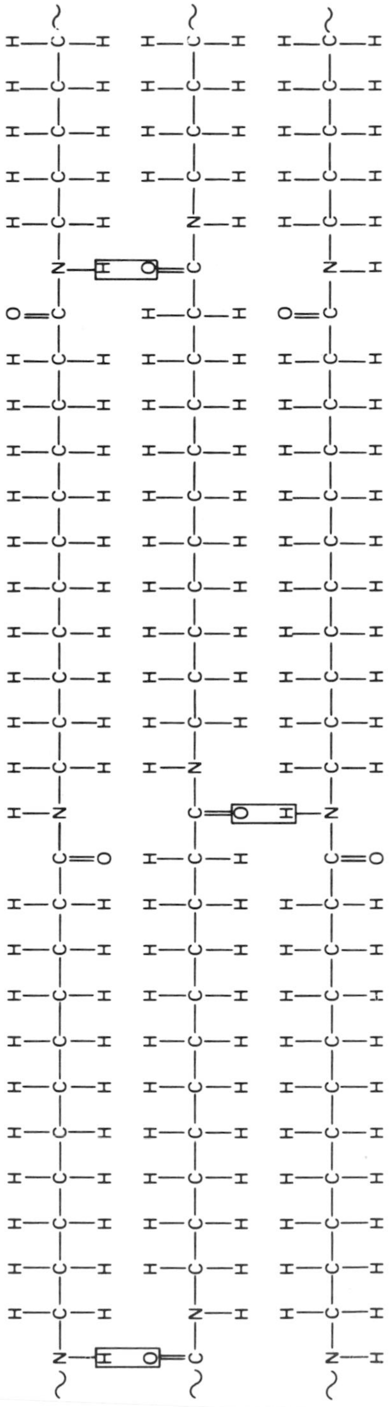

Figure 12.11 Intermolecular bonding of Nylon 11.

12.11, it can clearly be seen that regular ordered crystalline regions can occur through hydrogen bonding. Together with the high intermolecular attraction through the amide groups between adjacent chains, this structure provides strength to the polymer. In addition, hydrogen bonding provides a measure of flexibility in the amorphous regions. Hence the combination of high interchain attraction in the crystalline regions and flexibility in the amorphous regions gives rise to a strong, tough polymer.

Since the distance between the amide groups vary between the different polyamides, so will the properties. Those polyamides in which the distance is greatest and so the frequency of occurrance along the chain will be lowest, will have lower interchain attraction and so low strength, low melting point and will be softer. Where the hydrogen bonding is greatest the polyamide will have low water absorption, greatest chemical resistance and a high degree of crystallinity.

All polyamides are strong and tough with good flexibility, resistance to abrasion and dimensional stability. Unfortunately, due to their deterioration as moisture is absorbed, polyamides are only used for secondary electrical insulation applications.

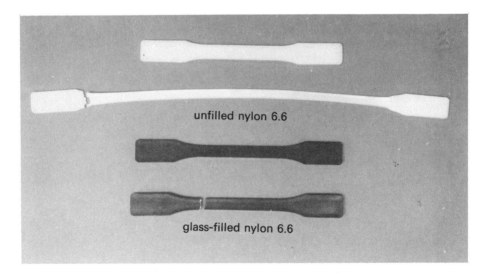

Figure 12.12 Tensile-tested specimens of Nylon 6.6 and glass-filled Nylon 6.6 to illustrate the difference in percentage elongation.

With further reference to diagrams of intermolecular bonding of the various polyamides, it can be seen that Nylon 11 has the greatest spacing between the amide groups followed by Nylon 6, then Nylon 6.10 with Nylon 6.6 having the least spacing or shortest *aliphatic segments* (CH$_2$ lengths). Considering crystallinity, the highest degree would occur where matching is nearest and where the aliphatic segments are shortest. The diagrams reveal that Nylon 6.6 is the best condition followed by Nylon 6.10, then Nylon 6 with Nylon 11 having the worst condition. It should be noted that if the polyamides are allowed to cool under

Typical properties of the various polyamides are shown in figure 12.13 for comparison.

Property	Nylon 6	Nylon 6.6		Nylon 6.10	Nylon 11	Copolymers	
		Unfilled	Glass-filled			66/610/6 40:30:30	66/610 35:65
Relative density	1.13	1.14	1.38	1.09	1.04	1.09	1.08
Yield strength (N/mm^2)	75	80	155	60	50	50	40
Elongation (%)	160	90	4	120	250	300	150

For most polyamides the maximum working temperature is of the order of 120°C

Figure 12.13 Properties of various polyamides

equilibrium conditions following processing, the molecular chains will align to produce the best-matched condition and so the highest degree of crystallinity.

The copolymers which are produced, Nylon 66/610/6 and Nylon 66/610 produce a more random arrangement of amide groups with less hydrogen bonding, providing greater flexibility at the expense of strength but with greater ease of processing.

In terms of water absorption capacity and ease of processing, Nylon 6 has the highest water-absorption capacity but is the easiest to process, whereas Nylon 11 has the lowest water-absorption capacity and is relatively easy to process. Nylon 6.6 is the most difficult to process with relatively high water-absorption capacity although Nylon 6.10 has a lower water-absorption capacity.

Most polyamides find application in the form of filaments and films with the exception of Nylon 6.6 which is used for moulding engineering components such as gears, bearings and cams and a variety of sterilisable medical and pharma-ceutical applications. The filament applications depend on the water absorption capacity and flexibility, the applications ranging from very fine brush tufting in Nylon 6.6 to fishing lines in Nylon 6.6/6.10. Nylon films also find wide application in pharmaceutical products and the packaging of foodstuffs, an example being the packaging of 'boil-in-the-bag' products. Other applications of the use of filaments are in glider tow ropes and as the tension members in composite belts for high-duty mechanical drives.

Chapter 13

Testing of materials

13.1 Reasons for testing

There are many reasons for carrying out tests on materials, the types of testing procedures which may be used are extremely varied, ranging from simple workshop methods of assessment carried out by semi-skilled personnel to sophisticated testing methods conducted by highly qualified technicians. Some of the reasons for testing materials are as follows:

(1) To check for imperfections within the structure of the specimens, e.g. cracks, cavities or inclusions. Such methods of examination may be part of a company's quality-control programme.

(2) To determine the mechanical properties of strength, hardness, toughness and ductility of the material. The results from such tests enable the designer to provide the most suitable shape and section capable of withstanding the expected stress levels.

(3) To evaluate the performance of a material in specific operating conditions. Such tests may be described as fatigue which involves alternating stresses, creep tests where the applied load remains constant or torsion tests where the operating stress is in the form of an applied torque. These tests may be carried out in environments which involve high or low temperatures, acidic or gaseous atmospheres or even various radiation levels.

As industrial production methods become more automated, some engineering materials will have to withstand continuous operating stresses throughout the functional life of the component. The need for improved material reliability together with energy-saving schemes to reduce fuel consumption emphasises the importance of reducing component weight whilst maintaining strength. The search for stronger more lightweight metals, alloys or composites will continue to be a major part of engineering research programmes. The methods of determining whether these 'new' materials will be suitable will be by conducting extensive mechanical tests. The results from such testing procedures will also highlight the limitations of certain materials, thus enabling the production of components which will function safely within the selected material's capabilities.

Of course, not all component materials require such detailed examinations of their properties. There are situations where very simple workshop assessment tests will be adequate. In such instances the experience of the engineer is relied upon to make a sound judgement concerning the material's suitability for the application in question.

13.2 Assessment of hardness

By definition, this property enables the material to resist indentation, abrasion or wear. A simple workshop test involves the use of a hand file on the hardened component. If the component is hard then the file teeth will make little impression on the metal. This degree or amount of impression can be used as a comparison of hardness between various materials. Alternatively a scriber point may be used in which case the ease with which the metal is scored will be an indication of hardness. Such testing methods depend for their accuracy upon experience and only serve to compare materials. An exact hardness value cannot be obtained by such methods.

These workshop methods can be used to check the efficiency of case-hardening, heat-treatment methods where the hardened skin, which may be 0.25 mm thick, is produced on the component.

13.3 Assessment of forming and ductility properties

Bend tests are normally used to test the quality of welded joints, they can also be used on certain forms of material which do not lend themselves to tensile testing. Such a test gives an indication of ductility which otherwise would be difficult to obtain.

Many sheet-metal operations require material to respond to bending and forming, without cracking around the bend areas. Such a property is described as bend ductility. Simple workshop tests can be conducted to assess this property. Figures 13.1 and 13.2 show simple bend tests in which solid straight pieces of material are subjected to pressure, and hence deformation in one direction. To be acceptable the material must be unbroken and free from visible signs of crack formation. In the automobile industry sheet metal, bent to produce wings, doors, bonnet panels etc., must have clean, smooth, crack-free surfaces for good paint

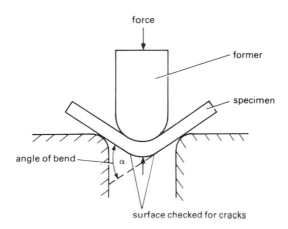

force

former

specimen

angle of bend

α

surface checked for cracks

Figure 13.1 Simple workshop bend test.

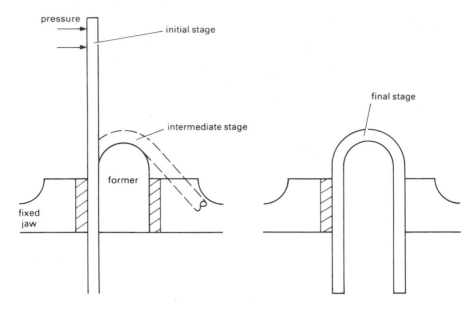

Figure 13.2 Simple bending operations.

finishes; the paint lustre would be marred by subsurface metal imperfections. Such imperfections could also weaken the component, since they could act as stress raisers. Bend tests give an indication of the manner in which the steel will behave when worked under forming rolls.

A workshop test for assessing bend ductility would incorporate a bench vice, and a forming mandrel. A number of specimens of equal cross section could be deformed individually to a specified angle and bend ductility determined from quality of surface finish. Similar bend ductility tests can be used on tubing and wire, and the bend ductility assessed by being wrapped around a specially prepared mandrel (figure 13.3).

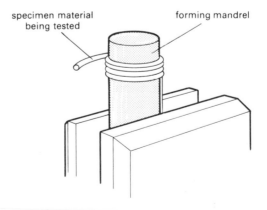

Figure 13.3 Assessing bend ductility.

For many forms of steel, particularly sheet and strip, it is desirable to test not only the capacity of the steel bend through 180° over a specified radius, but also its capacity to bend backwards and forwards through a considerable angle *without* developing a fracture. In such cases a reverse-bend test is used, where the strip is bent at right angles, first in one direction and then the other for as many reversals as may be necessary to break it.

13.4   Standardisation of material testing methods

Certain sections of the engineering industry, for example the aircraft industry, require very exact methods of material testing. Since a large number of components are required to complete an aircraft much of the material testing is conducted by other companies, under contract to the aircraft company. To ensure compatibility of results obtained from different testing laboratories a standardised technique has to be adopted. Such techniques involve standardisation of specimen dimensions, loading rates and values and the duration of the test. They are, in fact, covered by British Standard Specifications.

To ensure consistency and reliability of results when testing materials certain procedures must be followed. The specified testing methods may involve the use of specially prepared specimens, specific load values or even particular test duration times.

In hardness testing of materials, the most common technique involves the use of some form of indenter. These *indentation* methods assess the hardness of the material by the size of the indentation produced: a *hard* material will resist the indenter and produce a small impression, whereas a large impression indicates a soft material. If the load applied to the indenter is not strictly controlled then insufficient load will produce a small impression, whereas an excessive load will produce a large indentation. With hardness-testing machines that use a ball indenter, the ratio of indentation d to indenter D, i.e. d/D should range between 0.3 and 0.6. This ratio is achieved if the applied load P is related to the ball diameter D by the ratio P/D^2 which equals a constant for the particular material. The effect of incorrect load and hence incorrect interpretation of the hardness is shown in figure 13.4.

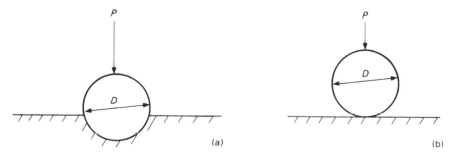

Figure 13.4 Influence of loading on depth of impression. (a) Excessive load, large indentation. (b) Insufficient load, very small indentation.

When hardness testing, the thickness of the material must be considered. Thin specimens can be subjected to excessive plastic flow resulting in distortion if the ball indenter and the applied load are too big. This is clearly shown in figure 13.5 which also illustrates the effect produced by the supporting platen which actually assists the resistance to indentation. Such hardness values are not to be relied upon. The recommended practice for soft materials is that the test piece should be at least 15 times the expected depth of the impression and for hard materials at least seven times the expected depth of impression.

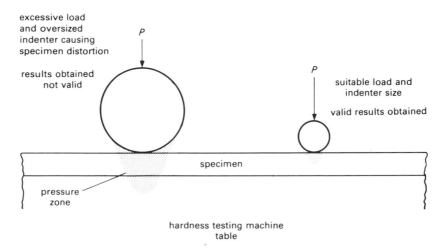

Figure 13.5 Importance of material thickness and depth of impression.

When the material to be tested is steel, then a constant is used. For steel this value is 30, and the following ratio P/D^2 is employed. For satisfactory testing of steel specimens $P/D^2 = 30$. Such a ratio enables various loads to be used and hence the correct ball indenter size can be determined. For a 10 mm ball, i.e. $D = 10$ mm, a load of 3000 kg would be used, likewise, a 5 mm ball would be used in conjunction with a 750 kg load.

When very hard materials are to be measured, the steel ball indenter has to be replaced by a diamond. This change is necessary because deformation of the ball indenter is likely to occur and therefore a reliable hardness value cannot be obtained. One suitable method is the Vickers Pyramid hardness test. The impressions produced are geometrically similar, irrespective of depth. This is achieved because the diamond indenter is ground to an angle of 136° between opposite faces and therefore the ratio of P/D^2 is not required (see figure 13.6).

To ensure reliable hardness values, the operator must consider various factors, including magnitude of load, material thickness, indenter size or type and indentation duration time, which is usually around 15 s.

The mechanical properties of tensile strength and ductility can be obtained by conducting tensile tests. The validity of the results depends upon certain factors, i.e. temperature, cross-sectional area, gauge length, surface finish, radii dimensions and the speed of testing. With most machines the tensile-test specimen may

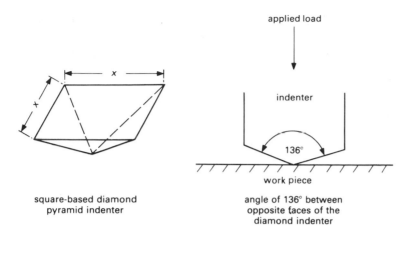

square-based diamond
pyramid indenter

angle of 136° between
opposite faces of the
diamond indenter

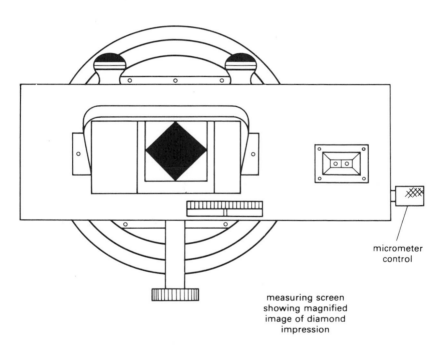

micrometer
control

measuring screen
showing magnified
image of diamond
impression

Figure 13.6 The diamond indenter.

be either round or of rectangular cross-section. Alternatively it may be cut from a plate or sheet. The chosen shape depends upon the nature of the product to be tested. The middle length or waist of the specimen is reduced in section or diameter to form the 'gauge length'. This ensures that the specimen will fracture within the gauge length. The dimensions and features are standardised to eliminate errors arising from variation in form. The specimens should comply

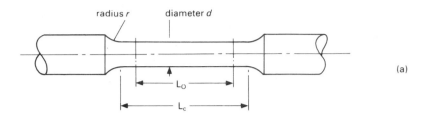

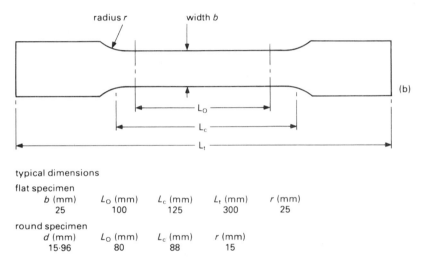

typical dimensions

flat specimen

b (mm)	L_O (mm)	L_c (mm)	L_t (mm)	r (mm)
25	100	125	300	25

round specimen

d (mm)	L_O (mm)	L_c (mm)	r (mm)
15·96	80	88	15

Figure 13.7 Typical dimensions of tensile-testing specimens. (a) Round and (b) flat test pieces.

with BS 18 (see figure 13.7 which illustrates the important dimensions for both flat and circular sections).

It is important to realise that the gauge length L_o is related to the cross-sectional area S_o by $L_o = 5.65 \times \sqrt{S_o}$. The radii shown in figure 13.7 should be closely adhered to; if they are too small then they may act as stress raisers which could result in premature failure. The quality of the surface finish should also be controlled in order to reduce the effects of stress raisers as should the speed of testing. The values of ultimate tensile strength and the yield point are affected by the rate of application of the stress during the test. An increase in the rate of loading brings about an apparent rise in the recorded tensile strength. A specimen which is stressed in tension, for prolonged periods, i.e. very slow rates of strain, may extend so much that it eventually fractures at a load well below the ultimate tensile strength of the specimen. The mechanical property of ductility is dependent upon the dimensions of the gauge length. If *any* value is selected that is not related by $L_o = 5.65 \times \sqrt{S_o}$, the value of strain, which is defined as extension/gauge length, and hence ductility, will vary for the same material! The ductility property is extremely important when forming operations are to be conducted. It is of prime importance that the above factors be closely followed when conducting tensile tests. The values of yield stress, tensile strength, elongation and reduction in area are determined on the basis of the *correct* dimensions.

13.5   Tensile testing of materials

When a material undergoes a tensile test, certain factors emerge concerning the way in which the specimen responds to the applied load. The type of fracture produced, i.e. crystalline or fibrous is an indication of the degree of ductility of the material. The speed of testing can also affect the way in which the specimen will fracture and the amount of extension produced before fracture. Whilst the tensile test is being carried out, a graphical record can be produced, the shape of which will be associated with the level of ductility of the material.

Consideration must be given to the factors which can affect the type of fracture of the specimen. Such factors may be the temperature of the specimen, the speed of testing, the quality of the specimen surface and its condition due to working or treatment. This type of information is extremely useful at the design stage since it provides feed-back concerning the suitability of the material. The degree of success of heat-treatment processes can also be assessed by the manner in which the specimens react to tensile loading.

By using tensile-test results, the design engineer can calculate the various levels of stress and strain that develop within the 'new' material during the test. The appropriate physical and mathematical relationships are outlined later in the text. Metallic and non-metallic materials are used in the design of many components that have to operate at various temperature levels. For such applications, testing at various temperatures is a method of determining the suitability of a material. However, should any test give unfavourable results then alternative materials for the particular application must be sought.

The tensile test is used for determining the strength and ductility of a material. The test involves an axial load being applied to specimens of circular or rectangular cross-section. The specimen dimensions are strictly controlled by relevant British Standards (see figure 13.7). The applied load is gradually increased at a constant rate and this causes the specimen to elongate and eventually fracture.

The amount of elongation and the type of fracture is affected by the type of material. Brittle materials show little plastic deformation before fracture (figure 13.8) whilst ductile materials show considerable amounts of plastic deformation (figure 13.9).

The fractured faces of brittle materials are crystalline and sharp and can easily be reassembled. Such a feature enables items made from china or glass to be easily repaired. Cast iron is a brittle material and figure 13.10 illustrates the negligible deformation that occurs during tensile testing.

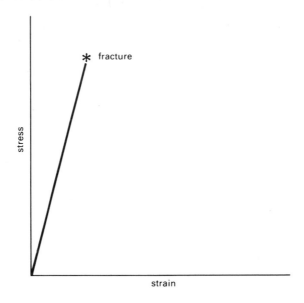

Figure 13.8 Stress–strain graph for brittle material.

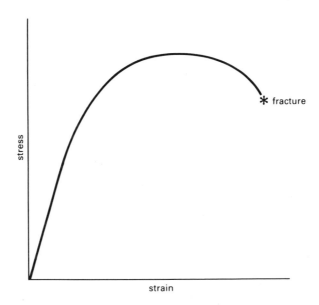

Figure 13.9 Stress–strain for ductile material.

With a ductile material, plastic flow results in the formation of a necked region together with a significant amount of extension (figure 13.11).

Graphical results of tensile tests will vary in shape, according to the structure or the treatment a material has received. The two important regions, the elastic and the plastic zones, are shown in figure 13.12(a).

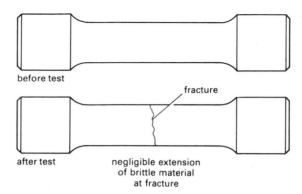

Figure 13.10 Brittle fracture.

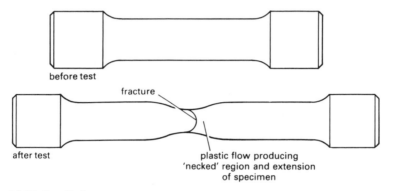

Figure 13.11 Ductile fracture.

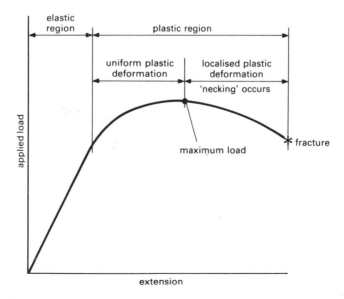

Figure 13.12(a) Typical tensile curve for a ductile metallic material.

The design engineer must be aware of the different material characteristics associated with these two zones. For springs or high-tensile bolts, the elastic region is of prime importance, whereas consideration must be given to the plastic region for forging operations. If during a forging operation the applied force remains within the elastic zone then no deformation will occur. Likewise, if the applied load to a spring creates plastic flow within the material, the spring will become ineffective, i.e. permanent change of shape will occur.

The tensile testing of materials is made more reliable, by utilising computer technology for control, graphical display and data processing. Curves of predicted results, can be superimposed on the actual graphical display during the test. All the characteristics of the material exhibited during the test can be stored by the computer for future reference and display. Two machines which incorporate computer systems are shown in figures 13.12(b) and 13.12(c).

The magnitude of the applied force to cause extension of the tensile specimen and the amount of elongation produced, depends upon the atomic 'make up' of the material under test. Some materials provide little atomic resistance and therefore slip easily, producing large amounts of elongation. Polyethylene polymer and annealed mild steel are such materials. In contrast, grey cast iron and thermo-setting polymers, described as *melamines*, have elongation percentage values below 1%. Apart from the structural atomic arrangements, affecting the

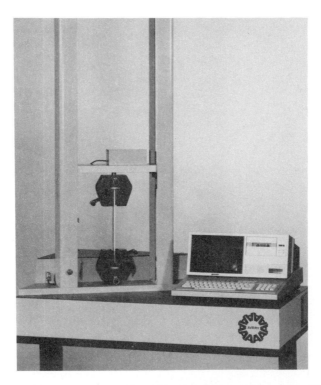

Figure 13.12(b) Tensile-testing machine with computer-control capacity 20 KN. (Avery-Dension Limited, Leeds).

Figure 13.12(c) The servo hydraulic universal testing machine 600 KN capacity. (Avery-Dension Limited, Leeds).

tensile and ductile properties, the temperature of the specimen and the rate of strain will also influence the shape of the load-extension curves. Mild-steel specimens at various temperatures are illustrated in figure 13.13.

The most ductile or 'plastic' materials are the ones that produce the largest percentage elongation. This property can be derived from the following formula:

$$\text{percentage elongation} = \frac{\text{final length} - \text{initial length}}{\text{initial length}} \times 100.$$

Associated with the increased specimen length, is a reduction in cross-sectional area, which can be determined from the following:

$$\text{percentage reduction in area} = \frac{\text{initial area} - \text{final area}}{\text{initial area}} \times 100.$$

The combination of percentage elongation and percentage reduction in area provides an accurate indication of ductility or plastic flow properties of a material. When a material provides resistance to elongation, within its elastic limit zone, it is considered a *stiff* material. This value of stiffness is described as the tensile modulus of a material. The higher the value of the modulus, the greater the force required to produce a predetermined amount of strain within the elastic

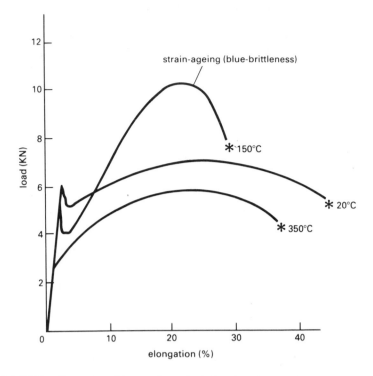

Figure 13.13 Tensile testing of low-carbon steel at various temperatures.

zone (figure 13.14). Mild steel has a tensile modulus of 200 GN/m², whilst aluminium has a value of only 70 GN/m².

Identical sections from these materials would indicate the superior stiffness of mild steel. In sharp contrast thermo-softening plastics such as polyethylene has a modulus value of only 1.4 GN/m² maximum, and it can be as low as 0.1 GN/m².

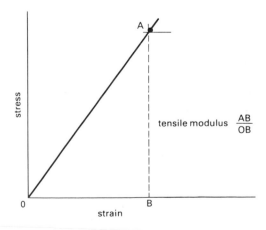

Figure 13.14 Graph to show the tensile modulus of a material.

Such polymers are not suitable for load-bearing situations; they would simply deform when subjected to stress. A more rigid polymer is PVC, with a tensile modulus of 3.1 GN/m^2. This level of stiffness makes it suitable for pipes, guttering etc. A comparison of various polymers and their tensile properties is shown in figure 13.15.

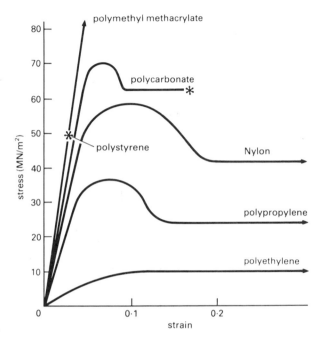

Figure 13.15 Tensile properties of various polymers.

The tensile results illustrated in figure 13.15 show a wide variation in properties, but all can be classified in terms of three basic characteristics. *Strong or weak* is assessed by the values of the tensile strength (maximum load/original cross-sectional area). The larger the tensile strength, the stronger the polymer. *Hard or soft* i.e. the higher the tensile modulus, the stiffer or 'harder' the polymer. *Tough-brittle* is associated with the area under the stress-strain curve. The smaller the area, the more brittle the material.

From figure 13.15 the polymers may be grouped as follows to emphasise their tensile properties: polyethylene, soft and weak; polypropylene, tough as polyethylene, but with an increased hardness value; nylon, strong and tough; polycarbonate (ABS), strong and tough; polymethyl-methacrylate (Perspex), hard and brittle; polystyrene, hard and brittle. The temperature of the specimen under test will affect the values of tensile strength and ductility.

It is sometimes required that a material has to function at elevated temperatures, e.g. high-temperature chemical-plant processes or heat-treatment equipment. In order to assess the suitability of materials, testing at elevated temperatures is undertaken. The effects of temperature on a particular polymer, polymethyl methacrylate, is shown in figure 13.16. From the graphical readings it

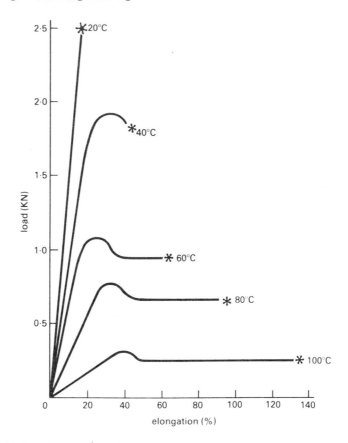

Figure 13.16 Tensile properties of polymethyl-methacrylate at various testing temperatures.

is shown that at above 60°C this polymer is tough and ductile, whilst below this temperature it is rather brittle. Metals are also tested at various temperatures. Typical results for low carbon steel are shown in figure 13.13. Information obtained from such graphs helps the designer-engineer to make the best possible selection of materials to suit the loading or environmental conditions.

The speed of tensile testing polymer specimens will also affect the results. Slow rates of testing enables uncoiling of the molecular chains resulting in gradual extension and a lower tensile-strength value. In contrast rapid testing prevents molecular slip from developing, thus producing a lower value of percentage elongation, but an increased tensile strength. Typical results are shown in figure 13.17.

The variation in speed of tensile testing metallic specimens, does not produce the same dramatic results as when testing polymers. Metals have a more ordered atomic arrangement; there is no uncoiling of the structure during testing but a certain degree of atomic slip or plastic flow will occur in the metal. To standardise the results obtained a British Standard gives guide lines concerning speed of testing, for polymers BS 2782, 1970, Parts 3/301 and 3/302, for metals BS 18, 1971, Part 2.

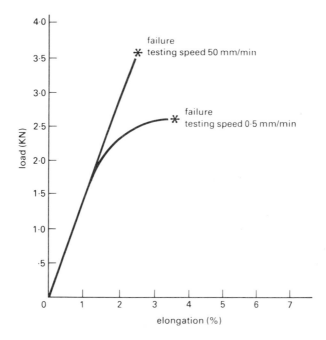

Figure 13.17 The effects of varying testing speeds on polymethyl-methacrylate.

Typical tensile-test results for cast irons, nylon and fibre-filled nylon are shown in figures 13.18 and 13.19. The various tensile curve shapes indicate that the tensile properties are related to the atomic arrangements within the materials. The grey cast iron (specimen 1) in figure 13.18 fractured before much plastic deformation. This is because the graphite flakes acted as internal stress concentrators and, a brittle fracture resulted. The malleable iron (specimen 2) indicated higher strength and ductility than grey iron. The alteration in properties resulted from the presence of nodular-shaped carbon or graphite. This modified graphite shape reduced the stress-raiser affects. With spheriodal cast iron the spheroids of graphite produced a good combination of tensile strength and ductility.

With the polymer specimens of nylon and nylon reinforced with glass fibre (figure 13.19) the presence of the glass fibre increased the strength by a factor of two. This increase in strength is at the expense of the percentage elongation.

Once results have been recorded graphically on metallic and non-metallic specimens, the following important property relationships can be derived.

(1)   E (Young's modulus of elasticity) $= \dfrac{\text{increase in stress}}{\text{increase in strain}}$

where stress $= \dfrac{\text{force}}{\text{original area}}$ and   strain $= \dfrac{\text{extension}}{\text{original length}}$.

(2)   Tensile strength $= \dfrac{\text{maximum force}}{\text{original area}}$.

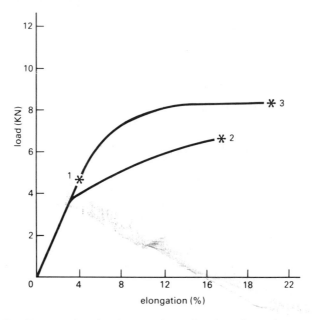

Figure 13.18 Tensile properties of various cast irons. Specimen 1: grey last steel. Specimen 2: malleable iron. Specimen 3: spheroidal graphite iron.

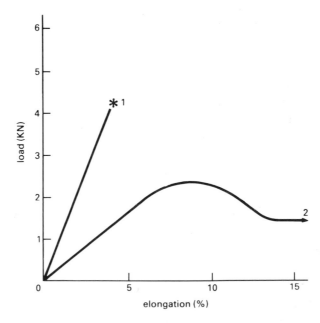

Figure 13.19 Tensile properties of nylon and reinforced nylon. Specimen 1: Nylon reinforced with glass fibre. Specimen 2: Nylon.

(3) Percentage elongation $= \dfrac{L_F - L_I}{L_I} \times 100$

where $L_F =$ final gauge length and $L_I =$ initial gauge length.

(4) Percentage reduction in area $= \dfrac{S_I - S_F}{S_I} \times 100$

where $S_I =$ initial cross-sectional area and $S_F =$ final cross-sectional area after fracture.

(5) Proof stress is required when the material does not display a clearly defined yield point. In design calculations the yield stress must not be exceeded. A well-defined yield point is shown in figure 13.20, whilst no detectable yield point is shown in figure 13.21. Typical materials are annealed copper and certain polymers. The process involves drawing a line parallel to the straight line portion of the curve. This line corresponds to an off-set distance between the lines of 0.1% of the gauge length (figure 13.21). The proof load is indicated by where the scribed off-set line cuts the curve. The proof stress is obtained by dividing the proof load by the original cross-sectional area.

(6) Secant modulus (Esc) is associated with certain polymers that curve from the origin (figure 13.22). Because no straight-line portion exists, a tensile modulus cannot be determined. The secant modulus is therefore used and is obtained by dividing the stress at a value of 0.2% strain by that value of strain, which is the slope of the line AB in figure 13.22.

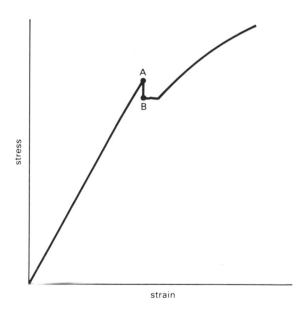

Figure 13.20 Stress-strain curve for mild steel showing sharp yield point. Position A: upper yield point. Position B: lower yield point.

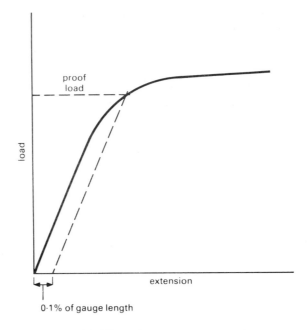

Figure 13.21 Determination of 0.1% proof stress.

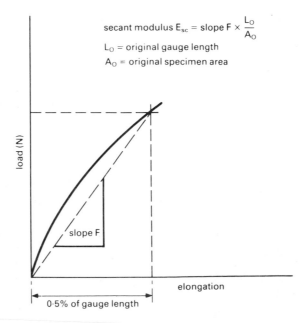

Figure 13.22 Determination secant modulus.

The following table indicates typical values of tensile strength for the most common engineering materials.

Material	Tensile strength (N/mm^2)
Grey cast iron	150–350
Copper alloys	200–1200
Zinc alloys	200–350
Mild steel	350–500
Nickel alloys	400–1600
Stainless steels	450–1300
Polyethylene	8–40
Polyvinylchloride	40–60
Nylon	70–90
Polystyrene	30–70

13.6   Torsion testing

The artistic ability of a blacksmith is clearly illustrated by the many curves and spirals he can produce in steel strip whilst manufacturing *wrought iron* components. The material used for such components offers little resistance to deformation by *twisting* and it is considered to have low torsional strength. In sharp contrast the design and manufacture of drills, boring bars and milling cutters requires a steel that possesses high-torsional-strength values. This property is usually achieved by a combination of alloying elements and appropriate heat treatment. Machine-tool components such as lead screws, feed shafts and arbor spindles also require high values of torsional strength.

The torsion test can be used in material-research programmes when determining the mechanical properties of new materials. The results, when analysed, will provide feed-back information on the materials capabilities. If necessary, alterations to composition or heat treatment can be undertaken, until the 'new structure' provides the desired results.

With the torsion test the specimen is gripped between two chucks. In this arrangement, one of the chucks is free to rotate whilst the other remains stationary. This allows a twist to be applied to the specimen, the extent of which is measured by the angular displacement of a point on the specimen relative to a scale.

If required, a torsion meter can be fixed to the specimen. The meter consists of two sleeves, which are free to rotate inside each other when the specimen is twisted, and are located by centring pins which *nip* the specimen at a distance equal to the gauge length of the specimen. A dial-test indicator is attached to one sleeve, while an adjustable arm attached to the second sleeve locates against the plunger of the dial-test indicator (see figures 13.23(a) and 13.23(b). Relative angular movement of the sleeves is indicated by the dial-test indicator to an accuracy of 1/10 000 radian. The torsion meter is used to obtain readings within the elastic limit of the material. When readings above the elastic limit are required, the angle of twist is read from a circular protractor scale.

13.7   Hardness measurement

With components that are designed to withstand high sliding or rolling stresses, e.g. ball or roller bearings, in order to maintain the precision of the bearings it is necessary to minimise wear. Such components will have to be hardened and the level of hardness will have to be accurately measured. The levels of hardness can only be verified by using accurate methods of measurement which may involve one or more of the techniques described below.

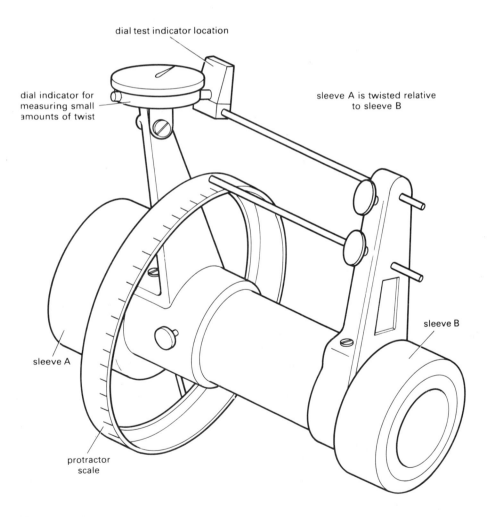

dial test indicator location

dial indicator for measuring small amounts of twist

sleeve A is twisted relative to sleeve B

sleeve B

sleeve A

protractor scale

Figure 13.23(a) Torsion meter.

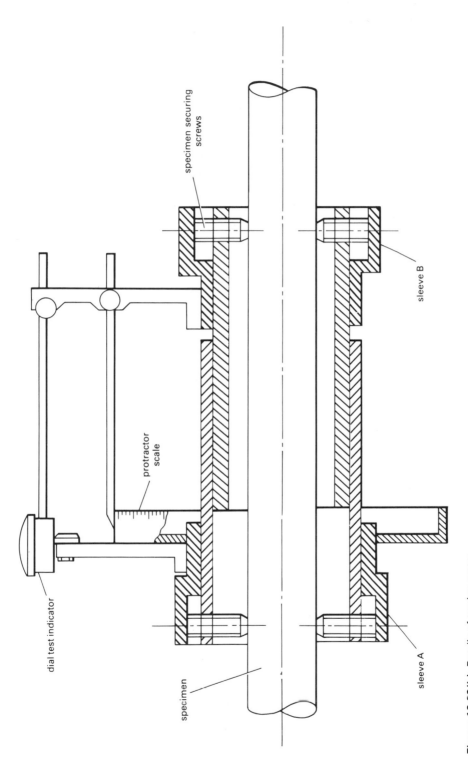

Figure 13.23(b) Details of torsion meter.

These methods may involve measuring the resistance to abrasion or scratching or to indentation. For measuring the resistance to abrasion a special scale called *Moh's hardness scale* is referred to. This scale comprises of a number of mineral substances which range from talc to diamond. The range of the scale is such that each substance can be scratched by the preceeding material, i.e. talc will be scratched by all the other substances, whilst diamond will not. The hardness number of the material under test is positioned one less than the substance that *just* scratches it. Moh's hardness test is suitable for materials that have negligible plastic flow properties such as ceramics, minerals and sintered components. The Moh's hardness numbers applicable to engineering practice are as follows:

Number	Mineral
1	Talc
2	Gypsum
3	Calcite
4	Fluorite
5	Apatite
6	Felspar
7	Quartz
8	Topaz
9	Corundum
10	Diamond

The following methods involve the measurement of the resistance of material to indentation.

Brinell test

This is carried out by pressing a hardened steel ball, under a known pressure, into a flat surface. The size of the ball most commonly used is 10 mm and the load corresponds to 3000-kg mass. The ball and load are related by a ratio $P/D^2 =$ constant, where the constant is dependent upon the material under test. The loaded ball sinks into the material for a period of 15 sec and produces a permanent spherical impression. This impression is measured using a Brinell microscope which is graduated in millimetres; tables are available to convert impression or indentation size to hardness values (figure 13.24). The hardness number BHN is found from the formula

$$BHN = \frac{P}{(\pi(D/2)) \, [D - \sqrt{D^2 - d^2})]}$$

where P = applied force in kilograms, D = ball indenter diameter in millimetres, d = diameter of impression in millimetres. A selection of indentation diameter and Brinell hardness numbers are shown in figure 13.24.

With the Brinell test it is important that a flat surface is produced on the specimen. If the surface to be tested is ground, then grinding must be done

BHN	Indentation diameter	BHN	Indentation diameter	BHN	Indentation
100	5.87	155	4.81	205	4.21
105	5.75	160	4.74	210	4.17
110	5.65	165	4.67	215	4.12
115	5.52	170	4.61	220	4.09
120	5.41	175	4.54	225	4.03
125	5.32	180	4.48	230	3.99
130	5.21	185	4.43	235	3.95
135	5.13	190	4.37	240	3.91
140	5.05	195	4.32	245	3.87
145	4.96	200	4.26	250	3.83
150	4.89	203	4.23	255	3.80

Figure 13.24 Brinell hardness numbers

carefully to avoid overheating the metal. If this precaution is not observed, some changes in the physical and mechanical properties may occur.

A power-operated Brinell-hardness testing machine is shown in figure 13.25.

Diamond pyramid test

This testing method uses a square-based diamond instead of a ball indenter; the diamond has an angle of 136° between opposite faces. The diamond indenter is forced into the surface of the specimen under an applied force. The applied load corresponds to various masses ranging from 5 to 120 kg. The usual load for hardness testing of steel is 30 kg. The diagonals of the indentation are measured and the readings converted to Vickers Hardness Numbers (VHN) by use of conversion tables, or it may be calculated as follows:

$$VHN = \frac{2 P_{sin}(\alpha/2)}{d^2}$$

where P = mass in kilograms, α = apex angle (136°), and d = average diagonal length in millimetres.

Since the indenter is a pyramid, the dimension across the indentation is proportional to the depth of penetration. Such a relationship enables hardness values of any material to be obtained directly, irrespective of applied force. Because the indentation is comparatively small, the surface should be polished to enhance and clearly define the indentation. This preparation improves the accuracy of indentation measurement.

Because of the small depth of penetration, it is well-suited to the testing of shallow hardness skins, for example nitriding treatment. The test is suitable for both high- or very-low-hardness readings, with a typical result for pure iron being 60 VHN, and certain carbides 2000 VHN. Figure 13.26 shows how the ideal diagonal value is obtained.

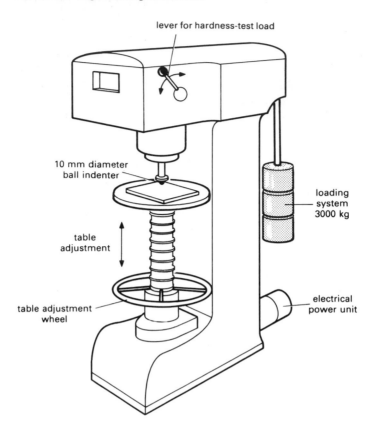

Figure 13.25 Brinell hardness-testing machine.

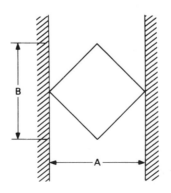

Figure 13.26 View through microscope eyepiece of diamond impression. Ideal diagonal value (A + B)/2.

The Rockwell test

This is a direct and very rapid method of determining the hardness of a material. No charts, tables or calculations are required. A number of different scales, each associated with a combination of load and indenter are needed to cover the range of hardness encountered in different steels. The scales most generally used are Rockwell B and C, these being used as standards on the measuring dial. Scale B is used with a 1.5 mm diameter hardened steel ball and a force due to a 100 kg mass. This is suitable for tests on mild steel, medium carbon steel and certain non-ferrous materials. Scale C employs a diamond cone having an angle of 120° between opposite faces and a 150 kg mass. It is used on hardened steel and most alloy steels. A machine for making Rockwell tests is illustrated in figure 13.27(a).

This hardness testing method is suitable for high-production processes taking only 10 s to conduct the test. Highly polished surfaces are not required.

The specimen is located below the indenter and the two brought into contact until the 'set' position on the dial has been reached. A primary load is then applied which eliminates errors due to deflection of machine and specimen. A secondary load is then applied which actually measures the resistance to indentation and the hardness value is indicated on the appropriate scale.

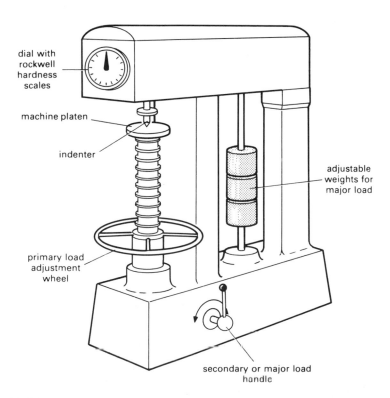

Figure 13.27(a) The Rockwell hardness-testing machine.

Figure 13.27(b) Digital testing for Rockwell hardness. (Avery-Denison Limited, Leeds).

With the development of micro-chip technology, hardness readings from Rockwell tests can now be illustrated in digital form. This method of hardness-value display, shown in figure 13.27(b) ensures a greater degree of accuracy and efficiency than the traditional machine illustrated in figure 13.27(a).

13.8   Impact testing

The toughness of a material is associated with its ability to absorb shock or impact loads without fracture. This property is required by the majority of engineering components, none more than in automobile construction. The engine, suspension and bumper units are frequently subjected to shock loads. These may arise from bad techniques, poor road surface conditions, or, the most common cause, failure to anticipate braking distances, thus producing the inevitable impact. An impact reaction forms the basis of toughness testing, the techniques of which will be discussed in conjunction with specimen design. The presence of a notch in the specimen will create a stress intensity in this region.

The test also reveals the material's tendency for brittleness which would not be detected by simple hardness or tensile-testing methods. The *notch effects* can be created by poor-quality, machined-surface finishes and by inferior weld runs, hence the importance of notch-impact testing methods.

Impact testing can be used to compare new materials with others which have proved satisfactory in service. Steels, like most body-centred metals and alloys absorb more energy when they fracture in a ductile manner than if they fracture in a brittle manner. The impact test is often used to assess the temperature of transition from ductile to brittle fracture which occurs as the temperature is lowered. The transition temperature is also dependent upon the shape of the notch in the specimen. For identical specimens, the sharper the notch, the higher the transition temperature. The fracture of the North Sea oil rig Sea Gem was attributed to a combination of low sea temperatures and notch effect resulting from poor weld runs. Hence impact values together with those of tensile and hardness tests should be considered when assessing the overall qualities of a material.

Many materials, including steel, can fracture in a brittle manner, i.e. very little plastic deformation taking place before failure. This brittle fracture is accentuated if the material is subjected to a rapid rate of loading, e.g. impact loading. The notched-bar impact test was devised to simulate these conditions, i.e. to test the resistance of a material to failure under the most unfavourable loading conditions. The principle of the test requires a notched specimen to be struck and fractured by a swinging pendulum. The quantity of kinetic energy of the moving

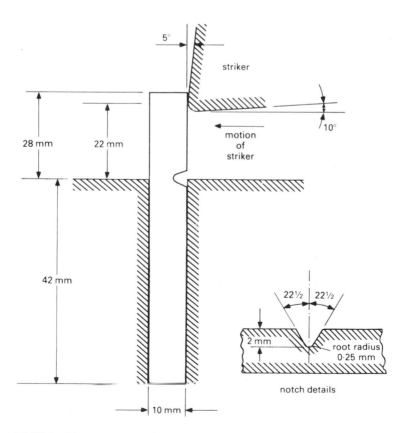

Figure 13.28 Izod impact test specimen.

pendulum which is absorbed to create fracture of the specimen is measured and can be related to the material's toughness value. The two most important notched impact tests are *Izod* and the *Charpy* tests.

The Izod test uses a notched specimen gripped at one end and held in a vertical plane. The specimen section containing the notch is struck by a swinging pendulum released from a datum position. Some of the kinetic energy of the moving pendulum is 'used up' during fracture. After fracture, the pendulum swings to a point past the specimen, the height of this swing determined by the energy absorbed during fracture. The weight of the pendulum multiplied by the difference in heights of swings is equal to the energy required to fracture the specimen, the units being newton metres (Nm) or joules (J). The impact test pieces have standard dimensions, see figure 13.28 for metals 10 mm square or 11.5 mm diameter. The Izod specimen may be provided with three 45° V notches, 2 mm deep arranged so that tests may be conducted along different directions in the material. The test piece for plastics are generally 12.7 mm square with a root radius of the notch being 1.0 mm, whilst for steel the root radius is 0.25 mm.

The pendulum has a strike energy of 160 J for metals and up to 20 J for plastics. The results from plastic specimens is expressed as the energy to fracture per unit width of notch (J/mm).

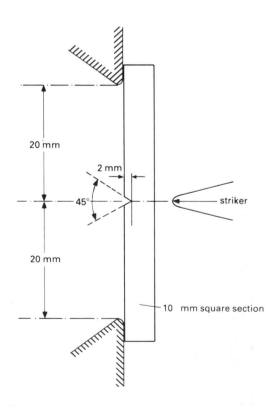

Figure 13.29(a) Charpy impact test.

The Charpy test uses a notched specimen that is supported as a simple beam (figure 13.29(a)) which is struck from behind the notch. The strike energy for metals is 320 J and up to 50 J for plastics. Metal specimens are usually 10 mm square with a V notch 45°, 2 mm deep and root radius 0.25 mm. Plastic specimens are 10 mm square and tested unnotched, with a milled slot or 45° V notch. The results are obtained in a similar manner to the Izod method and are expressed as Joules to require fracture for metal specimens, for plastics the results are expressed as energy, i.e. fracture per unit area of the cross-section J/mm^2. A Charpy impact-testing machine is illustrated in figure 13.29(b).

Bench-mounted Hounsfield machines will provide ideal impact results for metallic and polymer materials. For metallic materials two tups swing in opposite

Figure 13.29(b) The Charpy impact machine. (Special Steel Co. Ltd., Sheffield).

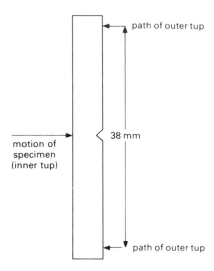

Figure 13.30 Specimen loading arrangement for Hounsfield impact test. Specimen of circular cross-section 7.8-mm diameter located in inner tup. Strike energy at impact 65 Joules.

directions, with one tup passing through the other. The specimen is held in the inner tup and is fractured as the tups pass. The strike energy involved being 65 J (figure 13.30). For polymers the specimen is placed on an anvil and fractured by a single tup. The loading is similar to the Charpy test, but on a much smaller scale.

Impact tests on specimens of varying carbon content at various temperatures are shown in figure 13.31.

Carbon content	Test temperature °C	Hounsfield impact value (J)
0.1	−40	50
	20	62
0.2	−40	18
	20	50
0.4	50	36
	−40	10
0.8	100	9
	0	6

Figure 13.31 Specimen results for Hounsfield machine

Chapter 14

Corrosion

14.1 Introduction

Those of us who own a motor vehicle will have experienced corrosion and will tend to believe that motor manufacturers design for corrosion rather than against it. However, the manufacturers deny this and take great pains to show how they try to combat the problem. Indeed, British Industry spends more than £2000 million each year on corrosion protection of ferrous metals.

Non-ferrous metals also corrode when exposed to the atmosphere. Some alloys exhibit visible signs such as the green oxide film on copper and the white powder deposit on aluminium alloys.

14.2 Theory of corrosion

Corrosion may be described as the result of a natural chemical reaction which restores a material back to its natural condition. This correctly suggests that all materials will eventually corrode, although corrosion is normally associated with metals.

The natural condition of most metallic materials is as an oxide. This is contained in the ore and requires refining to produce a useful metal. In the refining process, energy, in the form of heat is applied and useful metal is produced and can be said to contain a higher energy level than the original oxide since energy cannot be destroyed, it can only be converted from one form to another.

In corroding, as has been suggested, the metal reverts back to its original condition. In doing so, the energy absorbed during the refining process is gradually released. By observation, it is apparent that some metals corrode at a faster rate than others. Usually those requiring a higher energy input to refine, corrode at a faster rate. Hence corrosion of metallic materials can be associated with the release of energy, as the material regains an equilibrium condition.

Of all the forms of energy available, simple observations suggest that the energy released during corrosion is that of electrical energy. In addition, once corrosion has commenced, it is a continuous and irreversible process which further suggests that the electrical current flowing is direct rather than alternating.

Corrosion is a complex problem due to the many variables involved. The factors governing the rate of corrosion may be related to the metal itself and to the environment. The factors relating to the metal include: (a) the position of the metal in the electro-chemical series, (b) contact with dissimilar metals, (c) microstructure of the metals, (d) presence of internal stresses from cold working etc. The factors involved when considering the environment include (a) relative humidity, (b) presence of impurities in the atmosphere, (c) rate of supply and distribution of oxygen, (d) rate of flow of liquid, (e) acidity or alkalinity of the liquid, (f) presence of external stresses as a result of loading.

Corrosion takes one of two forms, depending on the conditions. The metal may react with gases in the atmosphere resulting in direct chemical attack—*oxidation*. Alternatively, the metal may react with aqueous solutions—*electrolytic corrosion*. These are the most common forms of corrosion.

14.3   Surface oxidation of metals

This form of corrosion attacks the surface of the metal and takes place in dry air. The ciemical action which takes place tends to produce an oxide film. In this reaction oxygen from the atmosphere combines with the metal at the surface to form a complex compound of the metallic oxide. The mechanism for this formation is simply described below.

Within a metal there are *free* electrons which, at the metal surface, are absorbed by oxygen from the air to form oxygen ions which then react with metallic ions to form the initial layer of oxide film on the surface (figure 14.1). After the initial layer of oxide has formed on the surface, further metallic ions and electrons diffuse through the oxide film to the surface. Gaseous oxygen at the surface is ionised by the absorption of the electrons which then reacts with the metallic ions. This forms more metallic oxide (figure 14.2). Hence, the rate at which the products of corrosion are formed depends upon the speed of diffusion of electrons through the oxide film. Since the flow of electrons depends on the electrical conductivity, as the thickness of the oxide film increases, the electrical resistance increases, thus slowing down the rate of electron diffusion and hence the ionisation of oxygen and so the rate of corrosion.

14.4   Electro-chemical corrosion

The most common form of corrosion is not surface oxidation—*dry corrosion*— but the more complex electrolytic action between metals in the presence of a

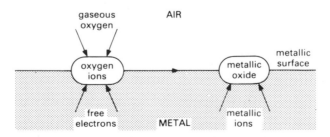

Figure 14.1 Initial formation of an oxide film (oxidation).

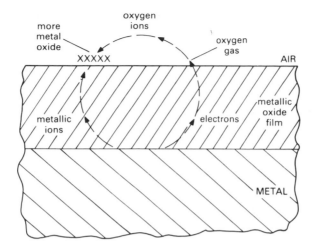

Figure 14.2 Further development of oxide film (oxidation).

moist atmosphere or in a liquid—*wet corrosion*. The mechanism of electrolytic corrosion can be better understood if there is an appreciation of elementary electro-chemistry.

Electro-chemistry

If two electrodes are connected to each other whilst immersed in an electrolyte, electrons can pass from one to another. The anode is the electrode to which negatively charged ions travel and the cathode is the electrode to which positively charged ions flow. Electron flow is always from the cathode to the anode.

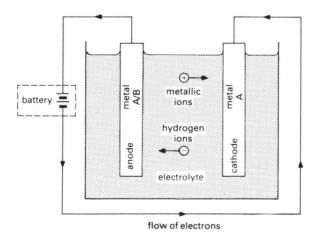

flow of electrons

Figure 14.3 Conventional representation of an electrolytic cell. Both anode and cathode of metal A with battery in circuit—electron flow. Both anode and cathode of metal A without battery in circuit—no electron flow. Anode of metal B cathode of metal A without battery in circuit—electron flow.

In an electrolytic cell, with electrodes of the same metal, a battery must be included in the circuit to create electron flow from the anode to the cathode. If there is no battery in the circuit, no electron flow will take place. However, if the electrodes are of dissimilar metals, electrons will flow from the anode to the cathode without a battery in the circuit. This effect is illustrated in figure 14.3.

A corrosion cell is simply an electrolytic cell where the process of electron flow proceeds spontaneously, the driving force behind the electron flow being the electrical potential difference between the anode and the cathode. The energy required is produced from the difference between the free energy in the metals of the electrodes. The flow of electrons is from the electrode of the higher free energy.

For corrosion to take place, the following conditions are required:

(1)   There must be a potential difference between the anode and the cathode. Hence, for metals, the anode and cathode must; (a) consist of different metals, (b) consist of different alloys of the same metal, and (c) have different concentrations of oxygen or electrolyte around them.

(2)   There must be an electrolyte present. Corrosion reactions are accelerated when the electrolyte consists of a salt solution. The presence of salt or impurities in the aqueous electrolyte increases the number of ions present, thus increasing the electrical conductivity. The ions are, in fact, the vehicle by which the electrons flow through the electrolyte and complete the electron circuit (figure 14.3). As the amount of salt increases, so does the flow of electrons through the electrolyte and hence the higher the rate of corrosion.

The pH value of the electrolyte is significant. The pH value of a solution is a number which expresses the concentration of hydrogen ions in the aqueous solution and thus is an indication of the degree of acidity or alkalinity of that solution.

The scale covers a range from 0 to 14, each number representing a definite degree of acidity or alkalinity. Neutrality is approximately 7, with values down to 0 indicating increasing acidity and values up to 14 indicating increasing alkalinity.

(3)   If the pH value of the electrolyte is less than 3, i.e. the solution is very acid, it is often possible for hydrogen to be evolved directly. If the pH value is in excess of 3, corrosion cannot take place by this mechanism.

(4)   An electrical connection must exist between the anode and the cathode to facilitate electron flow.

Hence *galvanic* corrosion takes place whenever two metals with different levels of free electrical energy are in physical contact with each other in the presence of an electrolyte. In this mechanism, illustrated in figure 14.4, hydrogen ions pass through the electrolyte from the low-energy cathode, react with the metallic ions produced from the high-energy anode and further react with free oxygen absorbed by the electrolyte from the air. This forms the hydrated product of corrosion whilst the flow of electrons proceeds from the high-energy anode to the low-energy cathode.

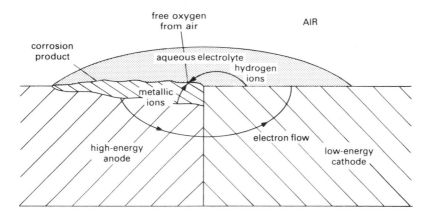

Figure 14.4 Mechanism of galvanic corrosion.

14.5 Potential difference in corrosion cells

In order to determine which of the two metals in a corrosion cell is likely to become the anode and so subsequently corrode, and which will become the cathode, reference may be made to the Standard Electrochemical Series. It can be seen from figure 14.5 that the series is related to the electrical potential of hydrogen.

The standard electrochemical series, however, is theoretical and deals with pure metals. In practice the same type of information is required for the more commonly used alloys. It is not necessary to know the precise electrical potential between a pair of metals, but merely to identify which one will corrode. In fact, the

	Metal	Electrical potential (mV)	
Protected end	Gold	+ 1680	(Cathodic)
	Silver	+ 800	
	Copper	+ 340	
	Hydrogen	0	(Reference)
	Lead	− 130	
	Tin	− 140	
	Nickel	− 250	
	Cadmium	− 400	
	Iron	− 440	
	Chromium	− 710	
	Zinc	− 760	
	Aluminium	− 1670	
Corroding end	Magnesium	− 2340	(Anodic)

Figure 14.5 Standard electromotive series

more widely spaced in the series, the greater the potential difference and so the greater the rate of corrosion. The summary of such information is contained in what may be described as the *galvanic* series as shown in figure 14.6. A useful extension to the galvanic series is a table (figure 14.7) which shows *corrosion tendencies* between couples which will cause corrosion and so which should be avoided.

	Series position	Material	
Protected end	1	Graphite	(Cathodic)
	2	Silver	
	3	Copper and its alloys	
	4	Nickel	
	5	Lead–tin alloys	
	6	Stainless steel (18:8)	
	7	Cadmium	
	8	Steels and cast irons	
	9	Chromium iron	
	10	Zinc and its alloys	
	11	Aluminium and its alloys	
Corroding end	12	Magnesium and its alloys	(Anodic)

Figure 14.6 A typical galvanic series

Anodic (corroding)	Cathodic (protected)
Iron (ferrite)	Cementite (combined carbon)
Iron	Graphite (free carbon)
Iron	Mill scale
Iron	Copper and its alloys
Iron	Nickel
Iron	Tin
Iron	Cadmium
Iron (non-aerated)	Iron (aerated)
Iron	Hydrogen
Iron (in stationary electrolyte)	Iron (in moving electrolyte)
Strained metal	Unstrained metal
Aluminium	Iron
Zinc	Iron
Magnesium	Iron

Figure 14.7 Common corrosion tendency couples

14.6 Differential aeration corrosion

The differential aeration cell is a typical concentration cell where the difference in potential between the anode and cathode is caused by the difference in concentration of oxygen around them. Oxygen starvation can be caused by localised liquid or solid on a larger piece of metal. A typical example is a water droplet on a metal plate. In this case, the area beneath the water is starved of oxygen and becomes anodic to the surrounding area (figure 14.8). With the water acting as the electrolyte and electron flow from anode to cathode, the hydrogen ions pass through the electrolyte from the cathodic area, react with the metallic ions produced from the anodic area and further react with the free oxygen absorbed from the atmosphere to form hydrated products of corrosion beneath the water droplet.

If oxygen starvation is created by a solid body, an electrolyte in the form of water vapour must be present before corrosion will take place.

Examples of differential oxidation corrosion can be seen in steel water-storage tanks, steel oil-storage tanks and steel pipelines transporting aqueous liquids, as shown in figures 14.9 to 14.11.

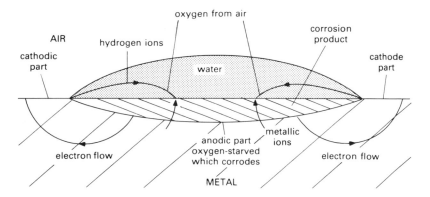

Figure 14.8 Differential aeration corrosion.

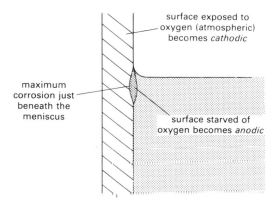

Figure 14.9 Waterline corrosion.

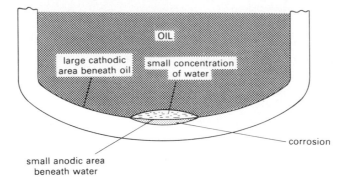

Figure 14.10 Corrosion at the bottom of oil tanks.

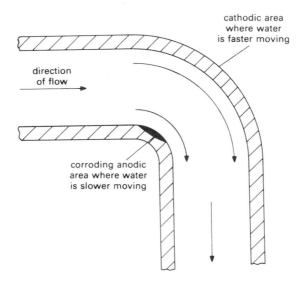

Figure 14.11 Corrosion at the inside of bends in pipes carrying a flowing liquid.

When studying corrosion, many factors must be considered. They include the composition and structure of the metal, the environment, the structural design and the operating conditions.

14.7   Composition and structure of the metal

Most metals comprise a solid solution which contain intermetallic particles. Although the particles vary in size and arrangement within the solid solution from one metal to another, invariably the electrical potential difference between the solid solution and the particles will be great, the solid solution being lower and hence corroding. Such is the case with ferrous metals— steels and cast irons.

In ferrous metals there exists both a solid solution—ferrite—and carbon-based compounds—iron carbide or graphite or both.

From the corrosion tendencies chart (figure 14.7) it can be seen that the carbon-based compounds are strongly cathodic to iron. Hence, when ferrous metal is in the presence of an electrolyte, the ferrite will corrode rapidly. This may

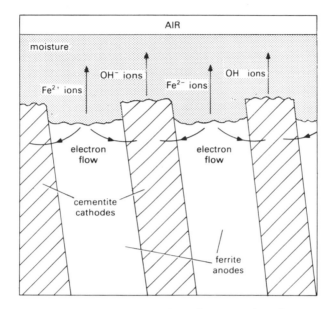

Figure 14.12 Electrolytic corrosion at the surface of a crystal of pearlite.

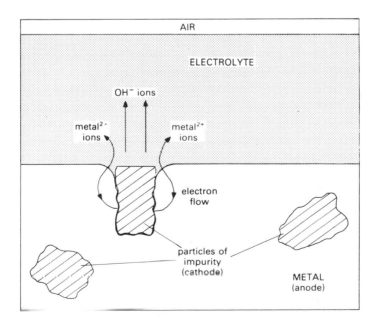

Figure 14.13 Electrolytic corrosion as a result of impurities in a metal at the surface.

also occur within a crystal, as is the case with pearlite (figure 14.12). Alternatively, if the carbon-based compound is more finely dispersed, as results from heat treatment, the corrosion resistance improves considerably. A similar situation will arise if impurities exist in the metallic structure. Again, the metal will be strongly anodic and will corrode in close proximity to the impurity (figure 14.13).

14.8   Environmental conditions

The principle environmental conditions which will influence corrosion result from impurities in the atmosphere.

Differential aeration and galvanic action induce an electrochemical potential difference which causes an electric current to flow. However, the rate at which corrosion will take place is governed by the rate at which the chemical reaction takes place. One of the most important influences is the nature of the atmosphere in contact with the steel.

The two atmospheric factors which give rise to an increase in the rate of corrosion of iron and steel are the high chloride content in the air in coastal areas and the high sulphite–sulphate content in the air in industrial areas. It has been found that the rate of corrosion in an atmosphere absent of either chlorides (which produce a hydrochloric acid electrolyte) or sulphates (which produce a sulphuric acid electrolyte) is in the order of 0.025 mm/year. However, in a marine atmosphere the rate can be double this value, rising to four times this value in a highly industrialised area.

Apart from the presence of sulphur and chloride gases in the atmosphere, there may also be dust. The corrosion from dust in the atmosphere is considerable and has a far greater effect than that of the sulphur and chloride gases. The dust particles provide the vehicle for ion flow through the electrolyte. The main components of air-borne dust are carbon, carbon compounds, metallic oxides and other mineral matter. It is the absorption of carbon dust particles in the sulphuric acid electrolyte which gives such a high rate of corrosion in highly industrialised areas.

14.9   Structural design

The design of structures should be such that there is no *permanent* contact of the metal surface with water. This is probably the most important method of protecting metal against corrosion.

Designs where water could be trapped without being able to evaporate readily are very prone to corrosion, even if the metal is well covered by a surface coating at these positions. The reason for this is that all paints and even plastic and rubber coatings are slightly porous and cannot be relied upon to exclude water and air absolutely. Once corrosion starts, even to a slight extent, the lift-off of the surface coating material by the more voluminous rust soon increases the rate of corrosion.

14.10 Stressing and temperature conditions

The stressed condition of a metal, whether internal or external will influence corrosion.

When a material has been cold-worked, the crystals are deformed—elongated in the direction of working. As a result of the way in which a metal deforms under such working conditions, a certain amount of the energy supplied to effect the cold working remains as locked-in residual energy. Hence a material in a cold-worked condition is in a higher state of energy than that which exists prior to working and as such allows metal atoms to escape the lattice more readily. It follows, therefore, that a strained material will be anodic to material which is unstrained and so will corrode more readily.

Similarly, when a material is subjected to external stresses from loading conditions, there will be areas of the material which are in a higher stress condition than others. Again, the more highly stressed areas will be anodic to those areas with a lower stress condition.

Corrosion is also affected by temperature. As a chemical reaction, its rate is increased with an increase in temperature. Scaling during heat treatment is an example.

14.11 Summary of corrosion prevention

Regardless of the complexity of the corrosion cell, there must be an anode and a cathode in electrical connection in the presence of an electrolyte. To prevent corrosion, this cell must be inhibited by rendering one or more factors ineffective. This can be achieved by taking one of the following steps: (a) make the electrolyte non-conducting, (b) break the electrical connection between the anode and cathode, (c) insulate either the anode or the cathode from the electrolyte, (d) remove the electrical potential from the cell.

14.12 Corrosion protection

There are many ways in which corrosion-attack of metals can be reduced, thereby protecting the metal against corrosion. They include the correct choice of the metal or alloy, cathodic protection, the use of protective coatings and the introduction of corrosion inhibitor chemical.

14.13 Corrosion-resistant alloys

The use of corrosion-resistant alloys is a widely-used method of providing corrosion protection although this may prove rather expensive, depending upon the particular circumstances. In such ferrous alloys the principle alloying elements are chromium and nickel, used individually or in combination.

When a ferrous metal containing a high percentage of chromium is exposed to an oxidising atmosphere an adherent passive film of chromic oxide (Cr_2O_3) is formed, preventing further penetration of oxygen and so further oxidation (corrosion) of the metal beneath it. If, however, the film is broken it immediately reforms in the oxidising atmosphere maintaining protection. It follows that as the chromium content increases so the corrosion protection improves.

When a ferrous metal contains a high percentage of nickel a solid-solution austenitic structure is produced in which all the carbon and other elements present are dissolved. This presents a uniform structure to an oxidising atmosphere and so considerably reduces the possibility of corrosion.

When nickel is present in combination with chromium, the nickel serves to improve the stability of the oxide film.

The corrosion resistance of other alloys can be improved by adding small amounts of particular elements, i.e. tin or aluminium to brass and magnesium to aluminium.

14.14 Cathodic protection

As stated in Section 14.5, there is a potential difference in a corrosion cell for which reference can be made to the Standard Electrochemical Series and in particular to the typical galvanic series (figure 14.6).

Where a metal requires protecting a galvanic couple can be formed by introducing another metal which has a lower electrical potential and so will be anodic to the original metal. In this case it is said that the original metal is protected by becoming the cathode in the corrosion cell—it is being given *cathodic protection*.

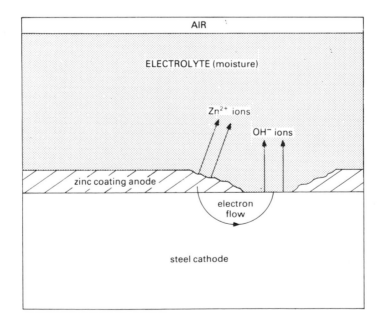

Figure 14.14 Sacrifical electrolytic corrosion at the position of a break in the zinc coating on galvanised steel sheet.

The protection is, in fact, sacrificial. Such is the case with zinc coating of steel. Initially the steel is protected by the coating. However, if the zinc coating is broken, the steel will still be protected sacrificially for as long as there is any zinc in the vicinity of the break (figure 14.14).

Sacrificial or cathodic protection may also be provided to the hull of a ship by securing blocks of metal of a lower electrical potential than steel below the water-line. In such cases the blocks of magnesium or zinc will have to be replaced from time to time.

14.15   Protective coatings

Corrosion attack can be reduced by the use of protective coatings. Protection in this case is purely mechanical, preventing atmospheric oxygen from chemically reacting with the metal at the surface. Of necessity the coating must be dense enough to exclude oxygen and durable enough to withstand the elements. It may be metallic or non-metallic. However, the most common form of coating is paint. The base or content can vary considerably. The choice of coating used depends upon the application, examples of which are as follows:

(1)   Metallic: electro-plating, hot dipping, metal spraying, condensation of metal vapours, metal cladding, cementation.

(2)   Non-metallic: oxides, phosphates, chromates.

(3)   Paints: metal-rich, oil-based, plastic-based, resin-based, rubber-based.

These coatings may be considered as permanent. However, temporary protection may also be given. Temporary protectives are often used to provide machined parts with protection during storage. In such cases the coating used must not cause any damage to the surface and must afford protection against moisture and oxygen. It must also be easily removed. Such temporary protectives may be divided into types according to their appearance. Basically they may be classified as soft films, including lanolin, hard films, including plasticised resins, greases and strippable plastic films. Whenever this temporary protection is used the metal surface must be perfectly clean and dry prior to application.

Another form of temporary protection is cocooning, originally developed for the protection of laid-up shipping but now widely used for the protection of machinery. The process consists of enclosing the machine in a sprayed film of impervious material with a dessicant (e.g. silica gel) placed inside it to absorb any trapped moisture.

14.16   Corrosion inhibitor chemicals

The use of vapour-phase inhibitors is widely used in industry. A vapour-phase inhibitor is a substance with a low vapour pressure whose vapour is inhibitive to the corrosion of ferrous metals. At normal temperatures the vapour keeps the enclosed metal parts free from rust. Some non-ferrous metals may, however, be corroded by the vapour. Temporary protection of small metal parts is achieved by wrapping them in paper impregnated with vapour-phase inhibitors or by including pieces of paper impregnated with these inhibitors in their packing. Their most common use is as paper bags impregnated with an inhibitor and with a vapour-barrier impregnation on the outside to keep the vapour in. This method is very effective for the storage of small tools, bearings, fastenings, etc. and has the advantage that the parts do not need cleaning prior to their use, as is the case with most other methods.

Chapter 15

Factors affecting material selection

15.1 Designer's considerations

During the design stage of a component, the engineer has to decide which material or materials would be most suitable. In certain circumstances the service requirements of the component dictate the materials to be used. For example, a soldering iron requires to have a copper *bit* in order that the heat energy can be conducted from the heat source to the joint—in this case a polymer bit would be unsuitable. In the majority of cases material selection is not as straightforward. Incorrect selection can result in component failure which could have disasterous repercussions. With the vast range of materials at their disposal, how do the design engineers make their final choice? The decision is not so simple and many important factors have to be considered including (a) size and shape of component, (b) cost of material, (c) availability of material, (d) number of components, (e) physical and mechanical properties, (f) dimensional tolerance, (g) availability of production plant.

The factors affecting the selection of a material cannot be considered in isolation. The size and shape of the component will influence the production processes, i.e. cast or machined. If the die-casting method of production were chosen, this would restrict the material selection to the low-melting alloys such as magnesium, zinc and aluminium. For die casting to be economical the component numbers would have to be high to counteract the high production costs associated with die manufacture. The degree of surface finish and the tolerances for the finished component will also affect the method of production— sand casting being inferior to pressure die casting with respect to both surface finish and dimensional accuracy, but it is cheaper. These are just a few of the interesting points to be considered by the design engineer.

15.2 Mechanical properties

The mechanical properties such as toughness, strength or hardness may indicate the suitability of certain materials for selection. Figure 15.1 illustrates the changes in the mechanical properties of iron–carbon alloys as the carbon content is varied. This provides the engineer with a range of iron–carbon alloys to select from. Once the appropriate alloy has been selected on the grounds of carbon content, will it function satisfactorily in service? One of the design requirements of small gear wheels is that they should be hard to ensure minimum wear, thus maintaining accuracy. Initially such gears were produced from hardened carbon steel but materials now used are *tufnol* or Nylon 6.6. Although the hardness

233

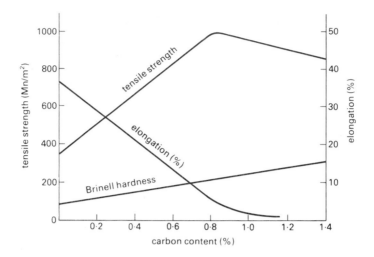

Figure 15.1 Relationship between carbon content and properties of iron–carbon alloys.

values of such polymer materials are less then carbon-steel alloys, in service the polymer gears perform satisfactorily, are quiet and do not require lubrication. The design engineer must be aware that specific properties possessed by a material do not reflect the way in which the material will behave in service.

The mechanical properties of a material can also dictate the method of production of the component. With some materials the properties of strength and toughness can render certain process methods difficult and therefore expensive. The extra cost factor could be due to increased manufacturing time or specialist tooling material or a combination of both. The solution to the production process problem could be by using the investment or lost-wax casting process, where the machining of material is not required. Careful consideration has to be given to the production process since this can have a direct bearing upon the mechanical properties, especially strength and toughness. If a component is produced by casting, it will have a particular crystal orientation and will not possess directional properties. If, however, the component is worked to shape by forging, then it will have directional properties. Worked metals or wrought structures are stronger than castings of the same material. The production of a small gear blank may be by blanking from hot-rolled plate, parting-off a thin section from a hot-rolled bar or parting-off a length from a hot-rolled bar and forging this into a blank. Although the parent material for the blanks possessed directional properties, individual blanks will have differing directional properties which could affect the mechanical property of strength. These directional properties are associated with *fibre* lines which flow with the crystal or grain deformation resulting from the direction of working.

15.3   Component cost

The cost of the component will be governed by the type of material selected and the process adopted for manufacture. Other invisible cost factors such as rates, rent and the general running costs of plant and equipment have to be considered before a final cost per component can be obtained. In many manufacturing industries it is considered that the purchase cost for material accounts for around 50% of the total manufacturing cost of the finished component. It seems, then, that the use of a cheaper raw material will reduce the overall cost thus making the component more financially attractive than its competitors. This, however, is not always true. When considering a suitable material for a subsonic airframe, a strong magnesium alloy involves fire precautions during machining together with subsequent corrosion protection. Titanium alloys are stronger and possess better corrosion resistance but are very expensive. Aluminium, however, is strong and relatively cheap. It is easily machined but it requires corrosion protection under certain conditions. From these three possible airframe materials, extra costs would be incurred in one form or another. Hence, a material which initially appears to be the cheapest is not necessarily so when all the production factors are considered.

15.4   Service requirements

The design engineer has to try to ensure that the material selected will remain stable and thus safe and will function in the manner it was primarily chosen for. In chemical- and nuclear-plant operations the major considerations are for structural stability, strength and corrosion resistance. Pure metals have extreme properties such as high corrosion resistance and very low strength but through alloying processes it is possible to maintain the corrosion qualities whilst developing the required strength. The various types of ferrous and non-ferrous alloys may provide the 'ideal' material for a particular service requirements. It may be that the weight of the component is critical, particularly in these times of energy conservation. Hence aircraft and automobile designs are very weight conscious in order to save fuel. The weight : strength ratio specification may be coupled with non-magnetic, corrosion-resistance or even creep or fatigue-resistance properties.

The designer must also be aware of material-treatment processes which will enhance the qualities of the various metals and alloys, thus ensuring that the service requirements are met. Such processes may involve simple case hardening, tempering or the more specialised treatments of austempering, nitriding or solution treatments. Because of the very varied demands made on metallic materials, a compromise may have to be made where the material chosen together with appropriate treatments which may be a surface treatment process such as anodising, case hardening or one which effects an internal structural change. Quench hardening or normalising may provide the structure that will cope with the service requirements.

It is often stated that all engineering problems can be solved by an appropriate method of design and whilst at the drawing board stage this seems to be the case. However, in practice many perfect designs fail miserably. This is often due to lack of experience and insight into the function requirements of the component material such as the unexpected stresses due to vibration, temperature variation or shock loading of the component.

15.5   Functional requirements

The material selected will have to function satisfactorily during its operating life. This will require it to be able to withstand all the expected loading situations. The loading situations or working environment can involve dynamic loading, static loading or hostile chemical atmospheres, as a result, various stresses are developed. When the component is subjected to a load that remains stationary it is described as a static stress situation. Components that have to withstand this type of load are legs on furniture, foundation structures of buildings, base casting of machine tools etc.

If the applied load is in motion (e.g. crane lifting equipment, vehicle suspension springs) or vibrational effects (e.g. pneumatic drills), then dynamic stresses are created. Such stresses are also created due to thermal effects of expansion and contraction. The design of heat-treatment equipment has to take into consideration this latter type of dynamic stressing. In chemical-plant design the stresses involved can range from static stresses associated with chemical-storage containers to stresses created by chemical attack, i.e. corrosion. This corrosive environment may be an indirect result of pollution produced by waste gases, or by direct contact with corrosive substances, i.e. salt solutions and acids. This working environment is very hostile, and expensive materials or surface treatments are required to slow down the corrosive attack. Whatever the rate of corrosion the component must remain strong enough to be safe and functional. Apart from obvious chemical environments stressing due to corrosive attack can be created by incorrect material selection. In such a situation an electrolitic cell is created where part of the component corrodes sacrificially. A reason why many structures and components fail prematurely arise from errors in judgement, concerning the capabilities of the chosen material and the actual stresses that are involved. With the development of computer technology and computer-aided design it is possible to assimilate the expected stress situations, thus improving the material selection process and component design.

15.6   Material properties

The success and expertise of the design engineer will be partly based on his or her knowledge of materials and how they flow or deform during the manufacturing process. Such an insight into the materials properties will ensure that the most suitable *forming* process will be adopted. Some materials such as lead cannot be successfully drawn into wire, nor does it respond to cutting-tool operations to produce the required form. Lead offers little resistance to applied force but it will not shear in a satisfactory manner to produce a quality surface finish. The property of malleability enables extrusion processes to be successfully conducted and hence production of lead wire. Although a simple and straightforward example concerning a material's reaction to different loading situations such as tensile (wire drawing), shear (cutting tool action) and compression (extrusion), it serves to illustrate the importance of the knowledge of a materials properties.

Different materials possess different physical properties. These properties will dictate material selection and are now considered in detail.

Tensile strength

This is the property of a material which enables it to resist the application of a tensile force. The atomic or molecular structure provides this internal resistance.

Hardness

The degree of resistance to indentation, abrasion and wear. Such reactions are achieved by heat treatment or alloying techniques, which minimise or prevent the atoms from slipping within the material.

Wear resistance

The ability of a material to maintain its physical dimensions when in sliding or rolling contact with a second member. It is associated with hardness.

Ductility

This property is associated with cold wire drawing operations and involves the gradual reduction in cross-section without rupture. Copper is very ductile but lead is not.

Impact strength

A measure of the response of a material to shock loading. Glass, grey cast iron and diamonds are materials with low impact strengths, whilst rubber, lead and certain polymers possess high strengths.

Electrical conductivity

A material that offers little resistance to the passage of an electric current is considered to possess a high electrical conductivity. It is the ease with which electrons will flow within a material when subjected to an external electron source. All metals and non-metallic materials such as carbon are good conductors of electricity.

Magnetic properties

Materials that are strongly attracted to a magnetic force field, produced electrically or by a permanent magnet, are described as ferromagnetic materials. Common elements that behave in this manner are iron, nickel and cobalt.

Thermal conductivity

The property of a material to conduct heat is known as its thermal conductivity. Certain materials such as copper and aluminium are excellent conductors. Because of this property such materials are used as soldering-iron bits, cooking utensils etc. In comparison asbestos, wood and certain polymers are poor conductors and are used as heat insulators, e.g. handles for cooking utensils, furnace gloves.

Density

The measure of a material to *pack* a certain mass into a given volume. A kilogram of feathers and a kilogram of lead have the same mass but the volumes are quite different. These materials have different values of density. It is an important factor where the weight and thus the mass are critical, e.g. aircraft structures. From a safety aspect the reduction of overall component mass must not result in reduced component strength.

Corrosion resistance

Materials that do not chemically break down in an alien environment are considered to have good corrosion resistant qualities. The 'alien' environment may be salt water moist air, polluted atmosphere or highly acidic atmospheres or solutions. The operating temperature can also affect this property, as can be observed by the oxide film that forms on hot steel. This reaction may be considered as accelerated corrosion or oxidation.

The physical and mechanical properties of a material selected for component manufacture will have to satisfy the fabrication requirements together with the service requirements. The ease with which the material will form, machine or become fluid for casting purposes will affect the final cost of the component. The reliability of the material whilst in service is also extremely important. Premature failure due to creep, fatigue or loss of hardness could prove to be costly if low-grade materials are used. The interaction of fabrication, service and economic requirements for ultimate material selection is shown in figure 15.2.

15.7   Material structures

A material such as a 0.4% carbon steel alloy that has undergone appropriate heat treatment can be used satisfactorily as a spring. Such an alloy is capable of withstanding shock loads without fracture, hence it is considered to be a tough material. Other materials that respond well to shock-loading situations are low-vulcanised rubber and many other thermoplastic polymers. Although all of these materials may satisfy the service requirements of toughness, it is only the alloys that gain this capability through heat treatment. In contrast the rubbers and polymers exhibit exceptional elastic properties in comparison to the alloys. The varied material reactions that arise in response to various treatments or applied

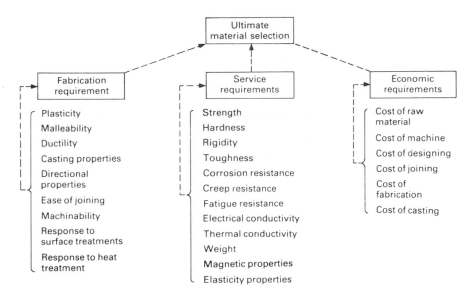

Figure 15.2 Factors that can control material selection.

stresses is due to the different internal structure of the metallic and non-metallic materials. Metals which can be distorted by working or heat treatment are of ordered atomic arrangements. These processes may restrict atomic slip, or, as in the tempering process which enables slip to take place more readily, the material becomes more plastic. In stark contrast rubber and the thermoplastic polymers are of entangled molecular chains; they do not possess the same degree of atomic arrangement as metals and hence cannot be described as crystalline materials.

The properties of engineering materials are associated with the internal arrangements of the atoms or molecules and how such arrangements interact with each other. The awareness of differences and similarities between materials is extremely important to the design engineer. Such knowledge enables the most suitable materials to be selected.

Materials can be grouped into categories depending upon the atomic or molecular arrangements; broadly speaking, engineering materials are either metallic or non-metallic.

Metals

If these materials contain ferrite or iron atoms they are classed as ferrous, if they do not they are described as non-ferrous. Such materials have their atoms arranged in regular patterns, called unit cells. The most common formations are face-centred, body-centred and close-packed-hexagonal.

Metallic materials have a high surface lustre, and possess good thermal and electrical conductivity. Such materials may be malleable, ductile, strong, tough or hard. These properties are governed by the amount of atomic slip occurring within the unit cell arrangements. Metals may be magnetic, non-magnetic and have varying levels of corrosion resistance. These materials respond favourably to either soldering, brazing or welding and can be shaped by forging, casting or machining.

Thermoplastic polymers

These materials have their atoms positioned to form linear or branched molecular chains (figure 15.3). Such plastics or polymers are capable of being repeatedly softened and reformed by the application of controlled heat and pressure. They are non-magnetic, and with the exception of certain acids possess good corrosion resistance. They can be formed by moulding, blow-vacuum forming and can be easily machined. Specialised fabrication methods enables the joining-welding of such polymers. Typical thermoplastic polymers are polyethylene, nylon and polyvinyl chloride (PVC).

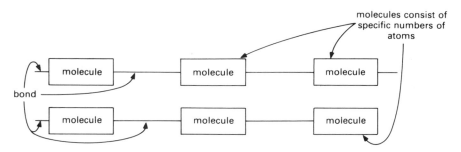

Figure 15.3 Linear molecular chain formation of thermorsoftening polymers.

Thermosetting polymers

These polymers are also described as thermohardening plastics. When such materials are subjected to heat and pressure a chemical change occurs, which produces cross links between the molecular chains (figure 15.4). It is the presence of these cross links that prevent the chains from sliding past each other, as a result the material cannot be reformed. Because of cross linking these polymers do not respond to the application of additives such as plasticisers and therefore a rigid inflexible material is obtained. Typical thermosetting plastics are phenolformaldehyde and polyurethane.

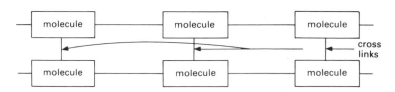

Figure 15.4 Formation of 'cross links' during curing that produces thermosetting polymer.

Elastomers

Certain polymer materials are capable of tremendous amounts of extension, without fracture occurring. One such polymer is natural rubber which can be stretched in an elastic manner to a considerable extent. Materials which respond by showing this type of elastic behaviour are described as elastomers.

Ceramics

These are considered as crystalline materials, i.e. their atoms are arranged in an orderly manner. They are very brittle and will fracture without yielding in a plastic manner. The use of ceramics range from china clay (kaolin) articles such as domestic crockery, electrical insulators, sparking plug insulation, to the production of drawing dies, cutting-tool tips etc. They possess higher hardness values than hardened carbon steel, and respond better to compressive rather than tensile stresses. The ceramic articles are shaped by moulding or pressing from the selected *clay mix*, and then fired or sintered to obtain the required hardness.

Wood

Such materials are extremely important to the construction engineering industry. They are divided into two distinct groups: soft wood and hard woods. Soft woods are obtained from spruce, pine and other cone-bearing trees. Hard woods are from deciduous trees, i.e. trees that lose their leaves every year. Common hard woods are oak, ash and elm. Before use the timber has to be seasoned to reduce its moisture content. For indoor applications the content is around 14%, whilst for outdoor uses it is around 22%. This material is liable to attack from fungi or insects therefore special preservative treatments are required, treatment can also provide fire-resistant qualities. The structure contains fibres or cells of long length, which are responsible for tensile strengths of around 6.2 MPa parallel to the fibres for hard woods, and 2.4 MPa for soft woods. Wood members can be fastened together by joints and glue or adhesives, they can be nailed, screwed or bolted. If loaded in a sensible manner wood is an extremely useful engineering material.

15.8 Material utilisation

Due to the enormous range of engineering uses for metals, polymers, ceramics and wood materials, the following list outlines the most-common engineering applications, the purpose of which is to illustrate to the student the possible scope and range of material utilisation. The section dealing with metals is divided into ferrous and non-ferrous, the ferrous being further divided into low carbon, high carbon and special alloy steels.

The engineer when designing a component has to select materials that will provide the required properties. When more than one property is required the selection of suitable materials becomes more involved. A comparison of

	Metals	Thermosetting	Thermoplastic	Elastomers	Ceramics	Wood
Density (kg/m³ × 10⁻³)	2–16	1.2–2.0	0.9–1.4	0.9	2–17	0.6 (Oak)
Hardness	Medium to high	Low to medium	Low	Low	Very high	Low to medium
Tensile strength (MN/m²)	Up to 2500	35–80 laminates up to 420	7–80	17–500	Up to 400	With grain 9 MPa, normal to grain 1.5 MPa
Thermal conductivity	Medium to high	Very low	Very low	Very low	Very low	Low
Corrosion resistance	Low to medium	High	High	High	High	Susceptible to fungal attack
Toughness	Low to medium	High	Very high	Very high	Very low	Depends upon grain direction
Electrical conductivity	Conductors	Insulators	Insulators	Insulators	Insulators	Insulators
Magnetic properties	High (except austenitic and non-ferrous materials)	None	None	None	None	None
Wear resistance	Low to high	Medium	Low	Low	Very high	Very low
Ductility (elongation (%))		0.5–1.5	2–850	700–1000	Negligible	Negligible

Figure 15.5 Comparison of properties.

properties is outlined in figure 15.5. At a glance the materials providing the necessary qualities can be isolated for more detailed analysis, until the most suitable available material or materials have been selected.

Ferrous metals

Class	Description	Application
Dead mild steel	Contains around 0.1% carbon	Automobile body panels, bright drawn sections
Mild steel	0.2%–0.3% carbon	Joists, shipbuilding boiler plate
Medium-carbon steel	0.4%–0.7% carbon	Springs, gears, axles
High-carbon steel	0.8%–1.2% carbon	Drills, lathe tools, files, taps and dies
Cast iron	2.3%–4.0% carbon grey cast iron malleable cast iron spheroidal cast iron	Engine cylinder blocks, brake drums Cam shafts, gears Crankshafts

Alloy steels

Low alloy steel	0.8% carbon 0.4% silicon 0.8% manganese	Valve springs, vehicle springs
Low alloy steel	1.0% carbon 0.45% manganese 1.4% chromium	Ball and roller bearings and races
Die steel	0.35% carbon 3.5% chromium 10% tungsten	Hot-working dies to operate at 700°C
High-speed steel	0.75% carbon 4.5% chromium 20% tungsten 10% cobalt 1% vanadium	Cutting-tool material to operate at high temperatures and severe pressures
Ferritic stainless steel	0.1% carbon 16% chromium	Capable of cold working to form sinks and other domestic appliances
Austenitic stainless steel	18% chromium 8% nickel	Due to excellent corrosion resistance used for chemical plant, food processing, nuclear-engineering components
Martensitic stainless steel	0.1% carbon 13% chromium 0.5% manganese	Cutlery, gas-turbine components

Non-ferrous metals

Copper	Refined	Where high electrical conductivity is required; good thermal conductivity and corrosion resistance, therefore used for radiators and boilers; used as an alloying element in brass and bronze

Non-ferrous metals (continued)

Zinc	Refined	Used for galvanising mild steel; an ideal die casting material for toys, automobile fittings; alloying element together with copper to form brass
Aluminium	Refined	Because of high thermal conductivity, good corrosion resistance and non-toxic qualities suitable for cooking utensils, food packaging, electrical conductors in grid system; alloyed with copper to produce Duralumin and used for aircraft components
Tin	Refined	Protective coatings on metals; important alloying element
Brass	Alloy	An alloy of copper and zinc, zinc content ranging from 10% to 42%
	10% zinc	Jewellery, shell cartridge cases, cold-worked
	30% zinc	shallow pressings, naval brass marine
	35% zinc	applications, high-tensile brass, marine
	37% zinc and tin	rudders, propellers
	42% zinc	
Bronze	Alloy	An alloy of copper and tin
	3% tin	Coinage, medal bronze, phosphor-bronze
	5% tin + 0.2% phosphorous	bearings for steam turbine, components for marine applications (Admiralty gun metal)
	10% tin + 2% zinc	

Thermoplastics (modern generic names given in brackets)

Cellulose nitrate (poly(cellulose nitrate))	Cutlery handles, tool handles, piano keys
Cellulose acetate (poly(cellulose ethanoate))	Artificial leather, toys, wire covering
Polyvinyl chloride (poly(chloroethene))	Rigid: piping, safety helmets Plasticised: raincoats, 'leather' cloth
Polyvinyl chloride acetate (poly(chloroethene) ethanoate)	Screens, protective clothing
Polyethylene (poly(ethene))	Packaging film, electrical insulation
Polypropylene (poly(propene))	Freezer components, cable insulation
Polystyrene (poly(phenylethene))	Toys, ceiling tiles, food containers, protective packaging materials
ABS (acrylonitrile butadiene styrene) (poly(propenenitrile) buta 1,3 diene poly(phenylethene))	Automobile bodywork, pumps, protective helmets
PTFE (polytetrafluoroethylene) (poly(tetrafluoroethene))	Bearings, non-stick coatings
Nylon (poly(amide))	Gears, fishing lines, brush bristles, ropes, cams, cable covering, clothes
Polyester (Terylene)	Clothing
Acrylics (Perspex) (poly(methyl 2-methyl propenoate))	Aircraft windows, lenses, baths

Thermosetting polymers (modern generic names given in brackets)
Phenolics

phenol formaldehyde (Bakelite) (poly(phenol methanal)	Electrical equipment, buttons, gears
urea formaldehyde (poly(carbamide methanal))	Adhesives, cups and saucers
melamine formaldehyde (poly(melamine methanal)	Building panels, crockery, electrical equipment
Polyesters	Laminated structures for boat hulls, car bodies, adhesives
Polyurethanes	Insulation foam, gears, upholstry packing, wire coatings, together with glass fibres for laminates

Elastomers

Natural rubber + 5% sulphur (vulcanised)	Vibration dampening pads, gloves, sealing rings, gaskets, sealing strips
Synthetic rubber	Footwear, hosepipes, cable insulation
Highly vulcanised rubber	Automobile tyres

Application of ceramic materials where qualities of refractoriness, wear, rigidity and strength are utilised by the engineering industry.

Ceramics

Ceramics for high temperature applications	Spark plugs, crucibles, furnace linings, pyrometer sheaths
Ceramics for strength and abrasion qualities	Sanitary wear, cutting-tool materials for rubber, plastics, some metals, (for continuous cutting action)
Ceramics for extreme temperature or abrasion qualities	
Ceramic material + metal powders cermets	Sintered tool tips, light-bulb filaments, rocket-engine components

Wood material

Soft woods	Pit props, paper pulp, hard-board, building industry requirements
Hard woods	Tool handles, foundry core boxes, foundry patterns, furniture
Laminated wood	Marine applications, furniture

Index

acrylonitrile 176
acrylonitrile-butadiene-styrene copolymer (ABS) 176
addition polymerisation 155, 169
age-hardening 112, 139
aliphatic segments 185
allotropy 54
alloying 11
aluminium 109
  alloys 109–120, 136
  bronze 144
amide groups of molecules 179
amorphous 162
annealing 69
  full 70–72
  process 31, 70, 71
  spheroidise 72
arrest point 10, 67
arsenic 135
arsenical copper 135
atmospheres—furnace 87, 88
atomic build-up 18
atomic slip 21, 63
austempering 79–81
austenite 54, 67
average molecular weight 162

bainite 79
batch furnace 84
bath
  lead 82
  salt 82
bend test 189
benzene 174
binary alloy 11, 31
binary equilibrium diagrams 33
body-centred cubic lattice (BCC) 14–16
bonding 3–8
branched molecular chains 157
brasses 142
Brinell test 210
brittleness 22
bronzes 143, 144

carbon 96, 152–155
carbon equivalent value (CEV) 98
cast iron 91
  grey 98
  inoculated 105
  malleable 101
  nodular 106
  pearlitic high duty 105
  spheroidal graphite (SG) 106
  white 99
cathode copper 134

cathodic protection 229, 230
cementite 44, 67
ceramics 241
cerium 106
Charpy test 216
chill crystals 26
chloro-ethylene 172
close-packed hexagonal lattice (CPH) 17
cohesive forces 2
cold shortness 44, 49
colourants 168
columnar crystals 26
complete solubility in the solid state 31
complete insolubility in the solid state 31
component cost 234–235
condensation polymerisation 155
constitutional diagrams 31
cooling rates 44
copolymers 156
copper 110, 134, 149
coring 46, 143
corrosion inhibitor chemicals 231, 232
corrosion prevention 229
corrosion protection 229
corrosion resistance 238
corrosion tendencies 224
covalent bond 4, 5–7
critical cooling rates 74
cross-linked molecular chains 157, 159
cross-linking 156
crystal 8
crystalline materials 30
crystalline structure 163
crystallinity 164
crystallites 163
cupro-nickels 135, 147
cyanide salts 89

decarburising reaction 87
deformation 24
dehydrochlorination 166
dendrites 18, 24
dendritic growth 18–21, 25
density 238
deoxidised copper 135
design considerations 233
diamond pyramid test 211, 212
differential aeration corrosion 225–226
diatomic 4
dilatometer 55
disordered zones 25
double bond 154
ductile cast iron 106
ductility 22, 189, 237
dry corrosion 220

elasticity 21
elastomers 21, 156, 157, 159, 241
electrical conductivity 237
electro-chemistry 221, 222
electrolytic corrosion 220
electrolytically refined copper 134
electrons 4
element segregation 47
EPNS 147
equilibrium conditions 31–32
equilibrium cooling 32
equilibrium diagrams 31, 33–34
eutectic point 35
eutectic reaction 35–36
eutectoid reaction 58

face-centred cubic lattice (FCC) 17
ferrite 55, 67
fillers 167–168
full annealing 70
functional requirements 236
furnace atmosphere 87–89
furnace
    batch 84–85
    muffle 84
    natural draught 85–86
    non-muffle 84–86
    tempering 86

galvanic corrosion 222
galvanic series 224
grain boundaries 24–27
grains 24–27
graphite 91
graphite rosettes 93
grey cast iron 98–99
gun metal 144

hardenability 76
hardness 23, 188, 237
hardness testing 191, 192
heat stabilisers 165–166
high density polyethylene 171–172
high polymers 162
homogenising anneal 49
homopolymers 156
hot shortness 44
hydrogen bonding 185
hypereutectoid steels 61
hypoeutectic irons 93–94
hypoeutectoid 61

impact strength 237
impact testing 215–218
indentation method 191
inoculated cast irons 105–107
inoculation 105
ionic bond 4, 5
iron 138

iron–carbon alloys 54–81
iron carbide 57
iron–graphite equilibrium diagram 92
iron–iron carbide equilibrium diagram 65–67
intercrystalline corrosion 142
Izod test 216

latent heat of fusion 1, 10
lead 138
lead bath 82
ledeburite 93
light alloys 108–133
limited solubility in the solid state 31
linear molecular chains 157–158
liquidus 35
low density polyethylene 170

macro-examination 27
magnesium 106, 110, 120, 124
magnesium alloys 127–131
magnetic properties 237
malleabilising 99
malleable cast iron 101–104
malleability 22
manganese 97, 110, 127, 138
martempering 79
material utilisation 241
martensite 44, 74
mass effects 76
mechanical properties 23
meehanite 105
mer 155
metallic bond 4, 7–8, 10
Moh's hardness scale 210
molybdenum disulphide 168
monomer 155
muffle furnace 84

natural ageing 112
natural draught furnace 85
natural rubber 174
neodymium 127
neutron 4
nickel 105, 138, 147
nickel–silvers 147
ni-tensyl 105
nitralloys 76
nitriding 76
nodular cast iron 106
non-muffle furnace 84
normalising 73
notch impact testing 214
nylon 168, 203

over-aged 112
oxidation 220
oxidising reaction 87

paired electrons   154
partial solubility in the solid state   31
pearlite   44, 58–60, 67
pearlitic high duty cast iorn   105
percentage elongation   199, 203
percentage reduction in area   199, 203
phenyl   174, 175
phosphorus   97, 135, 142
plastic flow   196
plasticisers   166–167
plasticity   21
polyamides   168, 177–187
polyethylene structure   166
polyethylene   169, 170–172
polymerisation—addition   155
polymerisation—condensation   155
polymers   152, 155
polystyrene   168, 174–177
polyvinyl chloride   165, 172–174
precipitation   41, 49
precipitation hardening/treatment   51–52, 109, 112, 127
process annealing   70–71
proof stress   205
protective coatings   231
proton   4
PVC   166, 168, 172

quench hardening   73–74
quenching
   media   88, 89
   procedure   89

rare earth metals   126, 127
reaction
   decarburising   87
   eutectic   35
   eutectoid   58
recrystallisation temperature   71
resistance to indentation   23
Rockwell test   213, 214
rubber   173, 174
ruling section   76

S-curve   69, 74
salt bath   82–83
secant modulus   205
semi-crystalline   163, 170
sensible heat   1
service requirements   235, 236
shortness—cold   44
shortness—hot   44
side appendages   157
side branches   158
silicon   96–97, 110
silver   124, 134
slip   23
solid solution   35

solidification process   9–11
solidus   35
solution treatment   127
solute   45
solvent   45
solvus   35
space lattice   11–12
spheroidal graphite cast iron (SG)   106
spheroidise annealing   72
stabilisers—heat, ultraviolet   165–166
Standard Electrochemical Series   223
states of matter   1–3
strength   21
styrene   174
styrene-butadiene rubber (SBR)   174, 175
styrene-cicylonitrite copolymers   175, 176
sulphur   97
super-cooled crystals   26
surface oxidation   220

temperature arrest point   10
tempering   75
tempering furnace   86
tensile test   195–207
tensile strength   203
theory of corrosion   219, 220
thermal conductivity   238
thermoplastic polymers   156, 161–164, 240
thermosetting polymers   156, 240
thorium   124
time temperature transformation curves (TTT)   77–78
tin   135, 138
tin bronze   143
torsion testing   207
tough pitch copper   135
toughness   22

ultraviolet stabilisers   165–166, 168
unit cell   12–13

valency electrons   4
van der Waal's forces   161, 164
vinyl benzene   174
viscous flow   158
vulcanising   158

water-absorption capacity   187
wear-resistance   237
wet corrosion   220
white cast iron   99
wood   241

Young's modulus of elasticity   203

zinc   110–111, 124, 127, 135, 138, 147
zirconium   126
zones—disordered   25